WITHDRAWN

The Laboratory Approach to Mathematics

The Laboratory Approach to Mathematics

Kenneth P. Kidd

Shirley S. Myers

David M. Cilley

Science Research Associates, Inc.
259 East Erie Street, Chicago, Illinois 60611

A Subsidiary of IBM

Science Research Associates, Inc.
259 East Erie Street
Chicago, Illinois 60611

Science Research Associates, Limited
Reading Road
Henley-on-Thames, Oxfordshire, RG9 1 EW England

Science Research Associates (Canada) Limited
44 Prince Andrew Place
Don Mills, Ontario, Canada

Science Research Associates Pty., Limited
81 Ryedale Road, West Ryde
Sydney, New South Wales, Australia

Permission has been granted by the following publishers for the
use of quoted material.

Pages 34, 35
Reprinted from *Mathematical Discovery*, Vol. 1, by George Polya.
Copyright © 1962 by John Wiley & Sons, Inc.

Page 126
Reprinted from "Some Theorems on Instruction Illustrated with
Reference to Mathematics" by Jerome S. Bruner in *The
Sixty-third Yearbook of the National Society for the Study of
Education.* Copyright, 1964, by Herman G. Richey, Secretary of
the Society.

Page 126
Reprinted from *The Developmental Psychologies of Jean Piaget
and Psychoanalysis* by Peter H. Wolff by permission of
International Universities Press, Inc. Copyright 1960 by
International Universities Press, Inc.

Page 128
From *Taxonomy of Educational Objectives*, Handbook II:
Affective Domain, by David Krathwohl, Benjamin S. Bloom, and
Bertram B. Mascia. Copyright © 1964 by David McKay
Company, Inc.

Page 158
Frank Riessman, "The Overlooked Positives of Disadvantaged
Groups," *Journal of Negro Education*, XXXIII (Summer 1964),
225–231.

Contents

Foreword

Active learning, student involvement, student participation, and relevance are educational concepts that receive the support of present-day educators. These concepts are consistent with the conclusions of educational psychologies and most present-day philosophies and are in tune with the inner yearning of the majority of our young people today. The laboratory approach to instruction embodies all these concepts.

The laboratory method of instruction as applied to the teaching of mathematics is not new. In 1906 J. W. A. Young wrote a mathematics methods text with a whole chapter devoted to the laboratory method.[1] Possibly the strong emphasis by Young on this method resulted from the efforts of John Perry in England and E. H. Moore in this country. Some quotations by these men should again remind us that a new idea in education is really quite rare and that educators of an earlier era did possess wisdom and an intuitive insight into the educative process. John Perry had this to say about education in his 1901 address to mathematicians in Glasgow:

> Now in my experience there is hardly any man who may not become a discoverer, an advancer of knowledge, and the earlier the age at which you give him chances of exercising his individuality the better. . . . Let him know that he is expected to be making discoveries all the time; not merely that the best established law is not complete, but that in the very simplest things it is not so much what

1. J. W. A. Young, *The Teaching of Mathematics in the Elementary and Secondary Schools* (New York: Longmans, Green, 1907).

he is told by a teacher, but what he discovers for himself, that is of real value to him, that becomes permanently part of his mental machinery. Educate through the experience already possessed by a boy; look at things from his point of view—that is, lead him to educate himself. I feel that throughout one's whole mathematical course it is important to teach a student through his own experiments, through concrete examples worked out by him.[2]

E. H. Moore, professor of mathematics at the University of Chicago, supported Perry's view in his address before the American Mathematical Society in 1902:

Would it not be possible for the children in the grades to be trained in the power of observation and experiment and reflection and deduction so that always their mathematics should be directly connected with matters of a thoroughly concrete character? . . .

This program of reform calls for the development of a thoroughgoing laboratory system of instruction in mathematics and physics, a principal purpose being as far as possible to develop on the part of every student that true spirit of research, and an appreciation, practical as well as theoretic, of the fundamental methods of science. . . .

Some hold that absolutely individual instruction is the ideal, and a laboratory method has sometimes been used for the purpose of attaining this ideal. The laboratory method has as one of its elements of great value the flexibility which permits students to be handled as individuals or in groups. The instructor utilizes all the experience and insight of the whole body of students. He arranges it so that the students consider that they are studying the subject itself, and not the words, either printed or oral, of any authority on the subject. And in this study they should be in closest cooperation with one another and with their instructor, who is in a desirable sense one of them and their leader.[3]

The point of view held by Perry and Moore certainly has a modern flavor. The above quotations could be used in chapter 1 of this book to support the importance of a laboratory approach in the present-day mathematics curriculum. There is now considerable research data to support the soundness of such an approach; in fact, it would be difficult for a teacher to argue against the use of the laboratory approach in the mathematics classroom at least part of the time. One might ask why more teachers during the past fifty to seventy-five years haven't used the laboratory approach in their classrooms. Undoubtedly an explanation could be made in terms of the demanding nature of the approach,

---

2. John Perry, "The Teaching of Mathematics," *Readings in the History of Mathematics Education* (Washington: National Council of Teachers of Mathematics, 1970), p. 226.

3. E. H. Moore, "On the Foundations of Mathematics," *Readings in the History of Mathematics Education* (Washington: National Council of Teachers of Mathematics, 1970), pp. 247, 250, 251.

the lack of background on the part of the teachers, the great influx of students into the secondary schools, the lack of interest of mathematicians in school mathematics programs, the sobering influences of two world wars and a depression, and the groping efforts made by psychologists in the area of psychology of learning. It is not important now to dwell on a rationalization for the failure of the Perry and Moore movement during the first half of this century. What is noteworthy is that there are many mathematicians, psychologists, and educators today (including Max Beberman, Edith Biggs, Jerome Bruner, Robert Davis, Z. P. Dienes, Harold Fawcett, George Polya, and W. W. Sawyer) who believe in instructional approaches consistent with those of Perry and Moore and are giving their time and energies to their implementation.

Will the laboratory method of teaching be adopted by a larger percentage of mathematics teachers during the next twenty-five years than during the first half of the century? There are reasons to believe so. Secondary mathematics teachers are generally better prepared and have more help from consultants and paraprofessionals. Publishing companies and supply houses are producing a wider variety of materials suitable for the laboratory. Also, the prevailing philosophies of education teachers and groups of teachers seem to imply at least some use of a laboratory approach to instruction and they seem willing to examine the approach with implementation in mind.

Before adopting a laboratory approach, teachers should have a fairly clear concept of what the method involves and should have given careful thought to and have tentative answers for questions such as the following:

1. What activities should be used in class?
2. When should a class demonstration be presented to the entire class and when should students work individually or in small groups?
3. How can a teacher provide instruction for everyone when there are many groups and individuals engaging in different activities?
4. How can a unit be planned so as to make maximum use of the laboratory approach?
5. What kind of curriculum materials should be available for student use?
6. What type of guidesheet can be used for small-group or individual instruction?
7. What is the role of evaluation and what should be the nature of reports to parents?
8. What type of facilities should be provided for a mathematics laboratory?
9. How can the approach be used so as to allow for individual differences among students?

The authors of *The Laboratory Approach to Mathematics* provide their answers to these questions, and their remarks should help teachers find answers that are unique to their own teaching situations.

The authors of this book have had a variety of experiences with the laboratory approach and are well qualified to write about it. Collectively they have set up laboratories in schools and have supervised teachers who use the approach; they themselves have used the approach in teaching elementary and junior high school students of various economic and achievement levels. Kenneth Kidd has been interested in the laboratory approach for nearly thirty years and probably has as many ideas on how to use the method at the secondary level as any educator today. In his classes he has been successful in helping future teachers become oriented to the laboratory method and in helping experienced teachers change their style of teaching to include more laboratory opportunities for their students.

The strongest aspects of this book, from a personal point of view, are the many examples and ideas that can be used in the classroom. Of special interest is the unit in Appendix B that starts a unit (Ratio) from the beginning and outlines it in detail and is consistent throughout with the laboratory approach. The book is excellent for mathematics education. Although the book was written for teachers of grades 5 through 9, many of the ideas can be used in secondary mathematics and university mathematics methods courses.

OSCAR SCHAAF
University of Oregon

Preface

The recent revolution in school mathematics seems to have lost its momentum. That revolution came at the end of the progressive period in American education.[1] It was started by the cries of Arthur Bestor,[2] Robert Hutchins,[3] and Paul Woodring,[4] and it was officially launched by Admiral Hyman G. Rickover,[5] who was outraged at the Russians' launching of the first space satellite. The editors of *Life*[6] and of *U.S. News & World Report*[7] added fuel to the engine, and from then until the late 1960s American educational reform, particularly in mathematics, was under way.

The revolution in mathematics education was mainly centered on curriculum reform; very little was said about method. The authors feel

---

1. Lawrence A. Cremin, *The Transformation of the School* (New York: Knopf, 1961).

2. Arthur E. Bestor, *Educational Wastelands: The Retreat from Learning in our Public Schools* (Urbana: Univ. of Illinois Press, 1953).

3. Robert M. Hutchins, *Conflict in Education in a Democratic Society* (New York: Harper & Row, 1953).

4. Paul Woodring, *Let's Talk Sense about Our Schools* (New York: McGraw-Hill, 1953).

5. Hyman G. Rickover, *Education and Freedom* (New York: Dutton, 1959).

6. *Life,* series on education: March 24 and 31, 1958; April 7, 14, and 21, 1958.

7. *U.S. News & World Report,* series of interviews with educators: Nov. 30, 1956; June 7, 1957; and Jan. 24, 1958.

that a new revolution is being launched and, though the launching is being carried out with considerably less fanfare, that it will have at least as great an impact on American education as did the former. The new revolution is in method, and there is much evidence (see chapters 1 and 5) to indicate that it will be along the lines indicated in this book.

The new revolution is sorely needed, for many students leave the schools practically illiterate in mathematics. These students are deficient in mathematical skills, concepts, and applications. We classify these students as low achievers in mathematics and many things have been said about them, but little has been done. Much of what has been said about educating such students is supportive of the laboratory approach (see chapter 6).

Our aims are modest. We wish to help teachers of grades 5 through 9 implement the laboratory approach. We hope to do this by providing them with a theoretical basis, with suggestions, and with specific examples. If it helps teachers of other than these grades, we will be overjoyed.

KPK   SSM   DMC

Acknowledgments

The authors wish to express their gratitude for the help of many persons: to Mossie Kidd, whose encouragement and typing helped spur completion of the project; to Pat Cilley and Edward Myers, whose patience and understanding allowed us to continue when further progress was difficult; to David, who gave many ideas their initial field test and who spent a number of hours helping get the manuscript ready for production; and to the many students past and present who have been inspired by the laboratory approach, and who have thereby inspired us.

We wish to thank the school officials who allowed us to work and photograph in their schools while the book was in production: Sister Mary Erna, principal of Holy Family Elementary and Intermediate School (Chicago, Ill.); Donald K. Cohen, principal of Charles Kozminski Elementary School (Chicago); James E. Hendee, principal of Locust Junior High School (Wilmette, Ill.); J. B. Hodges, director of the P. K. Yonge Laboratory School (Gainesville, Fla.); and Francis V. Lloyd, director of The Laboratory School of the University of Chicago. We are also grateful to all of the teachers who helped us so much, especially Sister Mary Monica Cahill, Sister Shirley Krzyzyk, Sister Yvonne Mattioli, Mrs. Mildred Williams, William Murphy, Raymond Lubway, Wayne Koch, and Mrs. Constance Smith.

We wish to express our appreciation to Frank W. McGraw of Gainesville for three of the fine photographs that appear in the book and to the publishers who granted us permission to use the quoted material.

Special thanks go to Miss Valerie Carlin of the SRA editorial staff for her fine editing and organizing and to Miss Suzanne Sumner of the SRA design staff for her magnificent design and untiring effort.

Finally, we wish to thank James F. Quinn, of Chicago, whose photographs do so much to complement and give meaning to the text.

About the Authors

Kenneth P. Kidd is the head of the mathematics education section of the University of Florida at Gainesville. His experience has included teaching junior high school, high school, undergraduate, and graduate students. He is a member of many professional organizations and has acted as speaker, consultant, and emissary for the National Council of Teachers of Mathematics (NCTM), for the Agency for International Development in India, and for many school districts throughout the United States. He has published articles in *Mathematics Teacher* and in *School Science and Mathematics;* as a member of the NCTM writing team, he helped write *Experiences in Mathematical Discovery.*

Dr. Kidd received his B.A. in mathematics from Maryville College in 1934, his M.A. in mathematics from Peabody College in 1937, and his Ph.D. from Peabody College in 1947. From 1939 to 1942, when Dr. Kidd was a U.S. Air Force captain in meteorology, he both attended and taught classes at the University of Chicago.

For thirty years Dr. Kidd has used the laboratory approach with junior high school students as well as with pre- and in-service teachers. In 1968 he made a study of the Nuffield project in England. As a member of the Committee on Mathematics for the Non-College-Bound, he had submitted a proposal for laboratory guidesheets to the committee. His proposal was approved by the board of directors of the NCTM in 1967. Dr. Kidd's commitment to the laboratory approach to teaching is deep and long-standing.

Mrs. Shirley S. Myers currently is a mathematics teacher at Oak Avenue Intermediate School in Temple City, California, a part time instructor at California State College (Los Angeles), and a visiting lecturer at Nazarene College. She has taught elementary school children, undergraduates majoring in education, teachers in in-service education, and persons taking adult education courses. A member of many professional organizations, she has acted as a speaker and consultant for the California State Department of Education, Los Angeles County offices of education, many school districts in California, the California Teachers Association and for the NCTM. She has published many articles in *The Bulletin,* a publication of the California Mathematics Council, and in *Common Denominator,* the publication of the Southern California section of the council.

Mrs. Myers received a B.A. from the University of California at Berkeley in 1944 and an M.A. in secondary education from California State College at Los Angeles in 1967. She has taken additional courses at the University of Connecticut, California State College (at the Long Beach and Los Angeles campuses), and at the University of California at Los Angeles.

In 1960 Mrs. Myers spearheaded the change of her school from one of self-contained classrooms to one of nongraded classes employing laboratory techniques. She also directs the manipulative materials center in the school. Her interest and involvement in implementing the laboratory approach in the classroom contributes practical knowledge and workable ideas to this field.

David M. Cilley is a staff editor in the Mathematics Laboratory at Science Research Associates, Chicago. Before joining SRA, he taught classes and supervised teachers at elementary, junior high school, high school, teacher in-service, and adult education levels. He belongs to many professional organizations and has been a speaker and consultant for the Illinois Council of Teachers of Mathematics, the NCTM, and SRA. At SRA he was one of the editors of the *Mathematics Structure and Skills* series and of the *Algebra Skills Kit.*

Mr. Cilley received his B.A. in mathematics from the University of New Hampshire in 1961 and his M.S. in mathematics from Fordham University in 1966; currently he is matriculating for a Ph.D. in mathematics education at Northwestern University. He has also attended C. W. Post College and the Illinois Institute of Technology.

Mr. Cilley used the laboratory approach for nine years with elementary, junior high, and high school students as well as with in-service teachers.

Involvement with the laboratory approach of these three outstanding educators totals more than fifty years. They have used the approach in classrooms from Gainesville, Florida, to Chicago, Illinois, and from Temple City, California, to Eastport, New York.

The *Laboratory Approach to Mathematics* is the first book to relate laboratory investigations to learning objectives, to place great emphasis on evaluative procedures for use with laboratory investigations, and to clearly state the rationale for the laboratory approach to teaching and to discovery learning. SRA believes that this book will be of great service to anyone interested in mathematics education.

1

Laboratory Investigations and the Curriculum

Never before has the mathematics teacher faced so many problems or had so many opportunities. The computer and other advances in science and technology are bringing about changes in the use of mathematics. The need for unskilled labor is disappearing; instead we are demanding increasing numbers of people possessing genuine insight into the structure of mathematics.

Recognizing this need, mathematics educators have made some important changes in school mathematics during the past decade. There has been a definite shift from an emphasis on manipulation of numbers and symbols to an increasing stress on the structure underlying these manipulations. For example, number systems with bases other than ten, expanded notation, and scientific notation are taught in order to provide insight into the decimal system of numeration.

More recently, many instructors have become dissatisfied with traditional teaching methods. They feel that a change in teaching method together with the changed curriculum can help each student realize his potential more fully. This search for a more appropriate teaching approach has led in two directions. One has been the development of ingenious material and equipment, such as film loops, overhead projectors, taped lessons, and learning kits, which are making the learning of mathematics more inviting for the student while enabling him to work and develop at his own pace. However, full use has not been made of this equipment, and many students are still not benefiting from it. The other direction in which this search has led is largely the result of the work of learning theorists. Their studies support the contention that students will learn concepts best if they are led to discover them themselves, through experiences related to the physical world. (Bruner [1968], p. 65; Dienes, pp. 17–42; Montessori, Introduction, pp. i–xxxix.) Consequently, instructors are becoming interested in a system of instruction referred to as the laboratory approach. It is a system based on active learning and focuses on the learning process rather than on the teaching process. It has produced exciting results in the teaching of almost all subjects and has been particularly successful in the teaching of mathematics.

Before discussing the laboratory approach in detail, we shall consider the nature of the subject matter and the goals of mathematics instruction. Let us begin by looking at the entire mathematical development of the students.

Nature of Mathematics

Mathematics is highly abstract. It is concerned with ideas rather than objects; with the manipulation of symbols rather than the manipulation of objects. It generally employs deduction rather than induction as a means of reaching conclusions.

Mathematics is a closely knit structure in which ideas are interrelated. Students who discover some of the structure of mathematics are often impressed by its beauty. They note the lack of contradiction, and they see how a new technique can be derived from one that has already been learned. Learning theorists use mathematics to study the ability of students to form concepts and discover patterns and interrelationships.

The generality of the abstract symbol is the underlying power of mathematics. However, it is also the basis for one of the greatest dangers in the teaching of mathematics—the danger that the teacher may overlook the importance of providing experiences that will make symbols meaningful to the students. Symbols facilitate thinking for some, but they impede it for those students to whom they are meaningless.

Models of the abstract ideas and symbols of mathematics are easily found in almost every aspect of our culture. Sets of objects may serve as models for the counting numbers; manipulations of sets of objects may serve as models of operations with counting numbers. A comparison of the number of turns of the rear wheel of a bicycle with the number of turns of the front sprocket wheel may serve as a model of ratio. Matching various lengths of a pendulum with the time of its swing provides a model of a function. A model of $\sqrt{2}$ is the diagonal of a square whose sides are each 1 unit long. The number line is an excellent model of the set of real numbers and is quite useful when introducing operations with negative as well as positive numbers.

Models can give meaning to the symbols of mathematics. Conversely, mathematical symbols can be used to answer questions about the properties of models.

Goals of Instruction in Mathematics

What goals should the mathematics teacher be striving to attain? A mathematics program is successful if the students can demonstrate the following at an appropriate level of difficulty:

1. *The ability to relate mathematical symbols and vocabulary to models.* Models enable students to give meaning to the abstract symbols and technical vocabulary of mathematics. For example, students may be helped to relate objects and events in their own environment to symbols and words such as the following: 8, π, 15%, 3, ratio, volume, function, and $A \cup B$. The student should be shown that mathematics is a logical ordering of the objects and events in the world; therefore the use of manipulative materials (models) and the application of mathematics to familiar things should help the student learn mathematics.

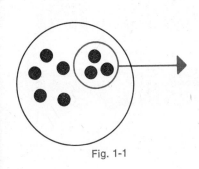

Fig. 1-1

The various operations on whole numbers can be related to the manipulation of sets of objects. The joining of two disjoint sets corresponds to addition of two numbers. A rectangular array of 5 rows with 7 objects in each row may be a model for 5 $\times$ 7. Subtraction corresponds to the process of finding one of the two addends in a problem like $\square + 3 = 8$ (that is, $8 - 3 = \square$). Division is the appropriate operation to describe the number of equivalent subsets into which a set of objects can be partitioned.

A: "$12 \div 3 = \square$" may mean "How many groups of 3 in 12?"

B: "$12 \div 3 = \square$" may mean "Distribute 12 objects into 3 groups of equal number. How many will there be in each of the 3 groups?"

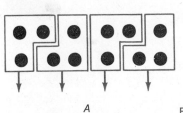
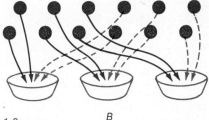

A   Fig. 1-2   B

The processes of mathematics can also be made more meaningful with the help of models. For example, a rectangular array of spots can be used as a model for 24 $\times$ 36. Since people don't generally memorize the products of all two-digit numbers, a simple numeral for the product

24 × 36 must be found by counting or by expressing each factor as the sum of tens and units and using the distributive law of multiplication over addition. The multiplication can be related to the manipulation of the spots in the array.

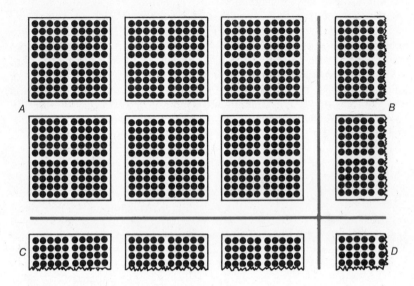

Fig. 1-3

For example:

$$24 = 20 + 4, \text{ and } 36 = 30 + 6.$$
$$\text{So } 24 \times 36 = (20 + 4)(30 + 6).$$

By using the distributive law, one obtains four partial products:

$$(20 + 4)(30 + 6) = 20 \times 30 + 20 \times 6 + 4 \times 30 + 4 \times 6$$

$$= 600 + 120 + 120 + 24$$

Then a simple numeral for their sum can easily be found as in the following.

```
        36                              36
     ×  24                           ×  24
     ────                            ────
        24 =  4 ×  6  (D)}              144
       120 =  4 × 30  (C)}
       120 = 20 ×  6  (B)}              720
       600 = 20 × 30  (A)}
     ────                            ────
       864                             864
```

2. *The ability to manipulate mathematical symbols.* The student should be able to work with symbols that represent ideas, thereby facilitating his thinking about these ideas. For example, suppose a housewife buys 2.73 pounds of hamburger at $.89 a pound. She suspects that the total cost printed on the package is wrong. What is the correct cost? In symbols this problem becomes $2.73 \times .89 = \square$. One who can use a calculator or is familiar with the numerical algorithm will be able to find the cost to be $2.43.

Filling out an income tax form is another familiar example of symbol manipulation facilitating thought. It can become an almost impossible process to comprehend if someone tries to see the whole process at once. Using familiar symbols for money and percents and manipulating these within a logical algorithm will allow the individual not only to find an answer to the problem, but also to understand what he is doing.

The concept of limits is extremely difficult for anyone who has not had a great deal of background in mathematics. However, it is possible to gain some understanding of this concept at a relatively early stage in the study of mathematics by working with familiar symbols, if the meanings of these symbols are clear.

Students should be given sufficient drill so that they will be able to manipulate symbols quickly and accurately.

3. *The ability to perceive the structure of mathematics.* Students may be helped to gain insight into the unifying ideas of mathematics. Students who understand mathematics see the relatedness of ideas; ideas are not learned as isolated facts. In the following examples, the multiplicative identity is the unifying idea.

$$\frac{2}{5} \div \frac{3}{4} = \frac{\frac{2}{5}}{\frac{3}{4}} \quad \text{and} \quad \frac{48}{64} = \frac{3 \times 16}{4 \times 16} \quad \text{and} \quad \frac{x^2 - x - 6}{x^2 + 4x - 21} = \frac{(x+2)(x-3)}{(x+7)(x-3)}$$

$$= \frac{\frac{2}{5}}{\frac{3}{4}} \times 1 \qquad\qquad = \frac{3}{4} \times \frac{16}{16} \qquad\qquad = \frac{x+2}{x+7} \cdot \frac{x-3}{x-3}$$

$$= \frac{\frac{2}{5}}{\frac{3}{4}} \times \frac{20}{20} \qquad\qquad = \frac{3}{4} \times 1 \qquad\qquad = \frac{x+2}{x+7} \cdot 1$$

$$= \frac{8}{15} \qquad\qquad\qquad = \frac{3}{4} \qquad\qquad\qquad = \frac{x+2}{x+7}$$

$$\text{(Assuming } x \neq {}^-7, x \neq 3)$$

4. *The ability to think creatively about mathematics.* To many people, mathematics appears to be a rigid system consisting only of symbols and a set of rules for manipulating them. Actually, mathematics has a great deal of room for creativity. In fact, the learning of this subject is greatly facilitated when students are challenged to use

Playing mathematical games helps
students think creatively about
mathematics.

their ingenuity to discover its many uses and properties. They should
be encouraged to give illustrations, formulate hypotheses, make
guesses, construct logical arguments, relate different mathematical
ideas, play games of strategy, work puzzles, and solve problems.

The games *Sprouts* and *Brussels Sprouts* (Martin Gardner, pp. 112–
15) can be used to help students see some of the more creative aspects
of mathematics. Each game involves topological invariance. After learning the rules and playing a few games, the children should be able to
begin to analyze the game to discover the strategy involved in it.

Another exercise that has possibilities for developing creativity is
that of guessing the number of beans in a jar. Various methods can be
used to arrive at a reasonable guess, and students should be encouraged to think of them themselves. For example, they could weigh a
small number of the beans, then weigh the whole jarful, and then empty
out the beans and weigh the empty jar; they could count the number
of beans in a small cup, count the number of cups of beans in the jar,
and multiply; or they could count the beans in one layer and count
the number of layers.

There are many additional activities and games that have creative
value. Students can be encouraged to formulate their own rule for
something, to test the rule, and to prove it; for example, when forming
simple polygons on geoboards students can be guided to discover a

Geoboards give students an
opportunity to formulate and test
a rule.

rule for predicting the area (A) of such polygons from a count of the number of pegs inside (K) the polygon and on the boundary (B) of the polygon (see guidesheet G-1, The Geoboard, in chapter 3). A proof of the formula $A = K + \frac{1}{2}B - 1$ can be found in an article by Niven and Zuckerman, in the December 1967 issue of the *American Mathematical Monthly* on pp. 1195–1200.

Students who can grasp flowcharting and the other rudiments of computer programming will have the basis for unlimited creative problem solving.

Students who are introduced to this aspect of mathematics will have a far greater appreciation for the subject than those to whom it is merely a dry, factual study.

5. *The development of favorable attitudes toward mathematics.* The atmosphere of the mathematics class should be such that the student comes to enjoy the study of mathematics. This kind of atmosphere would be one in which he gains confidence in his ability to use mathematics through experiences which are familiar to him and in which he can succeed. He should be led to see the need for precise language and for being able to defend one's conclusions, and to appreciate the significance of mathematics in the development of our culture.

Emphasizing the Goals of Instruction

When teaching any mathematical concept or process, the teacher must decide how much emphasis is to be given to each of these goals. Consider as an example the division of a whole number by a unit fraction:

$$6 \div \frac{1}{2} = \square$$

Should he go to the trouble of teaching the meaning of "$6 \div \frac{1}{2} = \square$"? If so, then he might first reinforce his students' understanding of division. He might start by giving the meaning of "$12 \div 3 = \square$" as "How many 3s are there in 12?" and then ask the students to relate this interpretation to "$6 \div \frac{1}{2} = \square$," or he might ask them to relate this equation to a model such as six candy bars cut into halves.

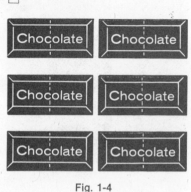

Fig. 1-4

Does he want to teach the students to manipulate symbols to replace the answer box with a number in "$6 \div \frac{1}{2} = \square$"? If so, he might want to develop one of the following procedures for manipulating symbols when dividing fractions:

Inversion rule	Multiplication by one

$$6 \div \frac{1}{2} = 6 \times \frac{2}{1} \qquad\qquad 6 \div \frac{1}{2} = \frac{6}{\frac{1}{2}}$$
$$= 12 \qquad\qquad\qquad = \frac{\frac{6}{1}}{\frac{1}{2}} \times \frac{2}{2}$$
$$= \frac{12}{1}$$
$$= 12$$

We may forget procedures if they are seldom practiced. In a recent interview with 50 adults chosen randomly in Gainesville, Florida, a university town, only 17 gave a correct answer to "$6 \div \frac{1}{2} = \square$." Most had forgotten how to work a problem of this type and were not able to give any meaning to "$6 \div \frac{1}{2}$."

Do we want to teach students why we manipulate symbols as we do? If so, we may want to justify the procedures above from the use of the field properties of the rational numbers.

Do we want to teach students to use "$6 \div \frac{1}{2} = \square$" to solve problems? If so, we may give prominence to the development of the ability to use the division of fractions in problem situations. For example, which of these situations can be solved by means of "$6 \div \frac{1}{2} = \square$"?

a) A certain recipe calls for 6 cups of sugar. How much sugar must Mary use if she is making $\frac{1}{2}$ of the recipe?

b) Mary has 6 yards of ribbon. She has enough ribbon for how many bows if each bow requires $\frac{1}{2}$ yard of ribbon?

Do we want students to find their work with fractions to be satisfying? If so, we may want to give attitude inventories that will help us determine which learning situations are satisfying and which should be revised.

Of the five goals, the second undoubtedly has received the most emphasis in the past. Textbooks were filled with symbols and rules for processing them. Students were required to spend most of their time practicing these processes. Students quite naturally learned to expect that their progress in mathematics would be based primarily on their ability to manipulate symbols accurately and quickly. This emphasis on the skills in manipulating mathematical symbols could have been justified only if it was assumed that by following through the prescribed steps in a procedure one would grasp the meaning involved.

$3\frac{1}{3}$ yards of ribbon cut into $\frac{2}{3}$-yard pieces—students see the meaning of $3\frac{1}{3} \div \frac{2}{3}$.

In recent curriculum developments more emphasis has been placed on structuring or organizing the ideas of mathematics. In general, however, teachers and authors are still attempting to reveal this organization to the students by expository methods. Too infrequently the students are helped to discover this organization by personal involvement. Even the integration and refinement of this knowledge is carried out by the teacher for the student rather than by the student for himself.

Today too little attention is being given to the other goals of instruction. Many students are not getting the type of experience with objects that would enable them to form meaningful mathematical concepts. Today, the youngsters' real world consists of such things as baseball, football, knitting, swimming, scouting, painting, Honda 90s, records, and automobiles. Seldom can we find textbooks and instruction in the mathematics classroom that are relevant to the students' experiences. Only recently have significant efforts been made to correct these shortcomings.

We also need to give more emphasis to the development of favorable attitudes toward mathematics. Research tells us that there is a positive correlation between a student's attitude and his achievement in mathematics (N.C.T.M., Twenty-first Yearbook, pp. 55–56). We must face the fact that an alarmingly large number of students possess unfavorable attitudes toward mathematics. They have a history of failure in this subject; they are threatened by mathematics and no one expects them to be successful in it. Their teachers, parents, and classmates expect them to fail; therefore they expect the same. Many of them are disorganized; they do little planning, actions are random, and concern is for immediate goals. It has been shown, however, that the attitudes of underachieving students can be improved and that such improvement will lead to improved performance (N.C.T.M., Twenty-first Yearbook, pp. 56–57).

The Laboratory Approach in Mathematics Instruction

We have indicated that in the past decade mathematics instruction has placed greater stress on the development of an understanding of the structure of mathematics. The next shift in emphasis in school mathematics may well be toward wider use of methods that will focus on the other goals of instruction, while teaching about the structure of mathematics in a more meaningful way. Concerned instructors have begun to investigate the use of mathematical instruction based on students' personal investigations and discoveries. They believe that investigations of the environment will do far more good than traditional teaching methods to build enthusiasm for and confidence in mathematics, to teach students to use their own ingenuity, and to relate mathematical ideas and symbols to real objects.

This type of instruction is referred to as the laboratory approach. The classroom in which it is best carried out is a room that has been reorganized and equipped to allow for individualized learning. Such a room is called a mathematics laboratory. This approach and special classroom have the following characteristics:

1. *Relates learning to past experiences and provides new experiences when needed.* The laboratory approach provides the student with new experiences to which he can relate the abstract mathematics of the classroom, and it helps him examine and organize his experiences with physical objects. Most teachers will agree that many students have an inadequate background from which to abstract mathematical concepts. For these students, suitable new experiences must be provided. For example, formulas for areas of regions may be meaningless to students who have not either covered such regions with square units or divided them into square units. Some students may have had many worthwhile experiences, but they need help in analyzing and organizing them so that mathematical ideas will make more sense. Much of their play contains hidden mathematics applications—balancing a teeter-totter by having the larger child sit closer, sharing four candy bars with five friends, figuring out how many minutes and how many seconds remain in a football game, figuring out how much cloth is needed in a sewing project, and so on.

2. *Provides interesting problems for the students to investigate.* A laboratory investigation usually starts with the presentation of a problem situation, which should be based on a subject of interest to the students. When possible, it should be stated as an open-ended question that will lead to discussion and to various channels of investigation. The problem may be stated by the textbook or by the teacher, or it may come from questions raised by the students. It can be investigated via teacher demonstration and class discussion involving the whole class, or by a small group of students or an individual working with a guidesheet (see chapter 3 and Appendix B). Using a guidesheet, the students can manipulate materials designed to help them answer questions that will help them discover an answer to the major question. The following are some examples of questions that can be investigated and the materials needed in addition to a guidesheet for the investigation: How are the various quantity measures related to one another? (Materials: cup measure, half-pint milk carton, pint milk carton, quart milk carton, half-gallon milk carton, gallon milk carton, and water, rice, or sand for pouring.) How does a slide rule work? (Materials for the elementary school: white strips of cardboard, rulers, and fine-tip black ballpoint pens; for the secondary school: all the preceding plus a table of logarithms.) How do you figure distances by referring to maps? (Materials: maps, globes, string, rulers, pencils, and paper.)

3. *Provides a nonthreatening atmosphere conducive to learning.* In the mathematics laboratory, students should feel free from threat of failure as they make predictions and try out ideas. The laboratory is a place for learning; it is not primarily a place to display a finished product removed from the false starts and fumbles of the investigators. The teacher no longer assumes the role of a fountainhead of knowledge

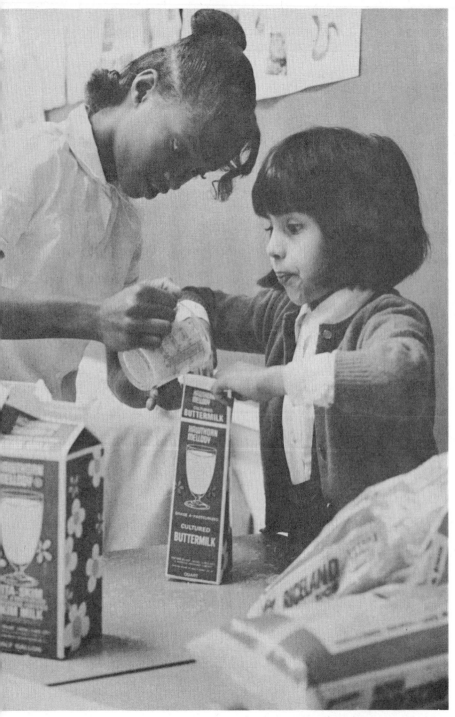

How many cups are in a quart?
A bag of rice can help you find out.

Students' questions provide a good source of problems to investigate in the mathematics laboratory.

Use of calculators will help ensure a failure-free atmosphere for the mathematics laboratory.

The laboratory is a place where you can work on your own.

Matching pendulum length to period —students discover functional relationships.

who reveals mathematics to his students. Instead, his role must be that of motivating, encouraging, and guiding students and of planning and supervising a variety of experiences for them. He must withhold generalizations and answers and allow students an opportunity to come to these conclusions on their own. He will soon realize that the majority of students can become curious about mathematical ideas. He will also realize that effective learning can take place in the classroom when curious students are seeking answers to questions. The French statesman Talleyrand is reported to have said, "What I have been told, I have forgotten; what I know is what I have guessed."

4. *Allows the student to take responsibility for his own learning and to progress at his own rate.* The laboratory approach takes into consideration the individuality of the learner rather than assuming that all students can learn through the same teaching approach. Each student has an opportunity to work at his own rate. He uses his own ingenuity in exploring the materials around him. He collects his own data and organizes it to solve a problem. For example, a student might collect data for the number of moves necessary to transfer given numbers of disks from one rod to another in the Tower of Hanoi puzzle and extrapolate to guess the number of moves for other numbers of disks; or a student might collect data related to the height a ball might bounce back after being dropped from different heights, then try to generalize his findings. This approach provides the opportunity for each student to assume some responsibility for his own learning.

The following are descriptions of actual classroom activities that illustrate the laboratory approach.

To make the introduction of functions to her algebra students as clear as possible, one teacher made use of a pendulum and stopwatch. A paper clip was fastened to a string hung from the end of a yardstick placed in the flag holder, and fishing sinkers were then attached to the clip. Ordered pairs of data were obtained—number of sinkers and number of seconds required for the pendulum of fixed length to make a swing. These pairs of data provided an example of a constant function. Next, by using 25, 50, 100, and 200 centimeters as the lengths of the pendulum with a constant weight, the students obtained another set of ordered pairs—measure of the pendulum length (l) in centimeters and measure of the period of the pendulum (p) in seconds. They then made a graph of the ordered pairs (l, p) and a graph of ($\sqrt{l}$, p). From these graphs they made a prediction of the period of the pendulum for lengths of 150 and 400 centimeters. These predictions were checked by experiment. The teacher then raised various questions. She asked what values might be used for l and why the points of the ordered pairs did not lie exactly on one smooth line. These experiences helped the students to grasp both the meaning and the significance of the function concept.

Occasionally a student will ask a question that the teacher cannot answer. This can be the basis for an interesting learning experience for both students and teacher. An example of this occurred when an

English teacher was asked to teach a class in eighth-grade mathematics. In class one day a student asked, "How could the Romans have used Roman numerals for multiplication and division?" Though the teacher did not know the answer, he was able to capitalize on this student's question to make the search for an answer one of the most exciting experiences in his teaching career.

Questions that arise outside the classroom will often lead to interesting discussions. One inquisitive seventh-grader asked, "Is it safer to make a trip from Miami to Chicago by plane or by automobile?" The student's family had talked about this because his father was planning such a trip. The entire class became involved in solving what they regarded as a significant problem. It became necessary to define terms and limit the problem to one in which data were available. The class found the following data for the preceding year: (a) the number of passenger miles flown on domestic commercial airlines; (b) the number of deaths on these flights; (c) the number of gallons of gasoline for which road tax had been collected; and (d) the number of deaths (of passengers as well as pedestrians) in automobile accidents. The class also had to find an average number of miles driven for each gallon of gasoline used as well as an average number of passengers in an automobile.

In 1949, students in the methods class of one of the authors entered into a project to measure the circumference of the earth. The class, which was in Gainesville, Florida, sought and received the cooperation of a college trigonometry class in Johnson City, Tennessee. The project was a modification of Eratosthenes' experiment of about 230 B.C. Two ten-foot poles were held vertical by means of a plumb bob, one at Gainesville and the other at Johnson City. The lengths of their shadows were measured on the same day, on level ground, when these shadows pointed to the north. From these shadow lengths the students were able to find the measures of the angles of elevation of the tops of the poles. The difference between these measures $(\alpha - \beta)$ represents the difference between the latitudes of Johnson City and Gainesville. The following proportion enabled the students to find C, the circumference of the earth: $(C : 360) = (m : \alpha - \beta)$, where m is the number of miles from Gainesville to Johnson City taken from a map.

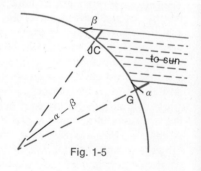

Fig. 1-5

Seven years later the same author met a young football coach while traveling to the Gator Bowl game in Jacksonville, Florida. The coach had attended East Tennessee State College in Johnson City. When he was asked if he had taken any mathematics while at ETSC, he said, "Yes, I took one year of college mathematics. There was one

Which are safer—cars or planes?
Many questions lead to interesting
mathematics problems.

Space perception is developed
through work with Soma cubes.

thing that happened in my trig class that I will never forget. Some man in Florida wrote to my professor and asked him to help measure the circumference of the earth."

Another teacher used the laboratory approach to help her students explore space perception. One team of students investigated the number of possible arrangements of four cubes such that each cube had a face-to-face contact with another cube. They had a box of cubes and a mirror (so that they could examine mirror images of each arrangement). Would the mirror image be the same arrangement or a different one? How would they define *different arrangements?* After finding the eight arrangements that are possible with four cubes, they decided to work with five cubes. This had not been planned by the teacher, and she did not know how many different arrangements were possible. At the same time, another group was working on a further extension of this activity that involved studying photographs of arrangements of cubes and trying to duplicate these arrangements. Others were working on making three orthographic views of each of several objects through the use of a "view box," making a one-point perspective view of a cube through the use of a clear plastic image plane, and identifying sections of objects from photographs.

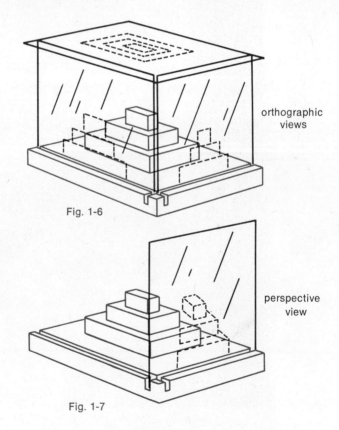

orthographic views

Fig. 1-6

perspective view

Fig. 1-7

A wooden block and three pieces of
clear plastic can help students to
draw orthographic views.

A rectangular array is a model that
can give meaning to multiplication.

Which shapes are rigid? You can
test them with wooden sticks.

Another teacher had his elementary school students working on a variety of activities to help them understand the operations of mathematics. Some were playing a game involving a balance scale designed to give physical meaning to multiplication. Others were working on the concept of division by cutting a measured length of ribbon into equal lengths. Several were investigating addition by using colored rods, while others were using sets of colored cardboard squares to clarify their concepts of addition and subtraction. Another student was building a concept of multiplication of large numbers by working with arrays of spots (10×10 per sheet).

A junior high class was grouped into small teams according to individual interest and need, to investigate several different areas of geometry. One group was working with volume relationships of geometric solids by pouring rice from one hollow model to another. A second group was working with geometry, functions, and combinatorics by drawing diagonals in plane convex polygons, keeping a record of the number of sides and the number of diagonals, and trying to derive from this a formula for the number of diagonals in an n-sided figure. A third group was investigating the property of rigidity as it applies to polygons of three or more sides. They constructed the polygons from sticks and tried to change the shape of each polygon while keeping one side rigid. After determining that they could change the shape of anything but a triangle, they repeated the experiment with a variation in which they found the minimum number of diagonals that would ensure rigidity.

Most teachers who are successful with the laboratory approach begin gradually. They usually make only partial use of this approach at first. The whole class does the same investigation or someone gives a demonstration to the whole class. As they become more comfortable with it, they gradually work toward its full use. The first four examples above are of such teachers. In the last three examples the teachers planned, scheduled, and supervised small teams of students who were observing and manipulating objects to find answers to questions. In these experiments all phases of the laboratory approach are illustrated.

Types of Activities Needed for the Mathematics Classroom

All classroom activities should be related to the instructional goals. It is important, therefore, for a teacher to begin with tentative long-range plans that are related both to these goals and to the scope and sequence of the course. The plans should take into account the competencies, attitudes, background, and habits of the students. Once the course outline has been made, the teacher must plan a variety of types of student activities, organized so that they will lead toward the achievement of the instructional goals. Activities should be designed to provide the following:

1. *Readiness.*   Some of these activities should be designed to provide experiences with objects and with specifics prior to the formalization of a mathematical concept at a higher level. Such activities, which we can call *readiness* activities, minimize the need for proficiency in symbol manipulation or in reading. In general, they consist of games, play activities, and open-ended questions leading students to explore and to brainstorm ideas with objects. Handling pie-shaped congruent pieces of cardboard circles is used by the first-grade teacher as readiness for the formal and symbolic treatment of fractions in the third grade. Using frames as placeholders is good background for the study of open number sentences. Filling boxes with congruent blocks is a readiness activity for youngsters who will later develop formulas for volume. Working with geoboards and with sticks pegged together will prepare students for a more systematic analysis of the properties of polygons. Readiness activities may also be provided in the elementary grades for concepts such as ordered pair, function, and square root, which will not be formally developed until a later stage.

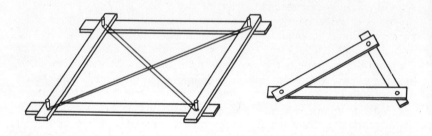

Fig. 1-8

2. *Concept Development.*   Other activities should be designed to lead directly to the development of specific concepts. Developing concepts involves concentrating on selected experiences and organizing these into a structure. Activities designed for concept development must be more carefully selected and more carefully structured than readiness activities. Dialogue between teacher and student can help the student select and organize his experiences. For example, an elementary school student might measure various parts of different-sized circles to develop the concept of π. The teacher's suggestions and questions or a guidesheet will help him to concentrate on circumferences and diameters. Further guidance can help him discover that the ratio of these two measures is always about 3.14 to 1 regardless of the size of the circles. Students in junior high school might be led to investigate the ways that squares, sticks, and cubes of wood can be put together to form rectangles and squares. This activity, with proper guidance from the teacher or a guidesheet, can help the student learn how to factor polynomials.

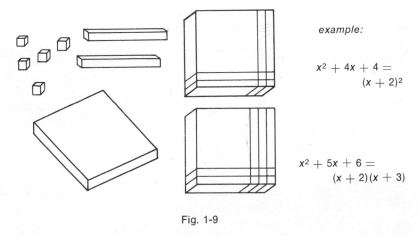

example:

$x^2 + 4x + 4 =$
$(x + 2)^2$

$x^2 + 5x + 6 =$
$(x + 2)(x + 3)$

Fig. 1-9

3. *Concept Synthesis.* A third type of activity should allow the student to review, organize, and integrate mathematical ideas. Activities of this type could include teacher-directed discussions, student discussions, student preparation of a display or oral report, and student reading of the textbook or a paper prepared by the teacher. For example, the students who took part in the activity involving the discovery of π just described could make posters and present their results to the teacher or to the class. Also, the students who performed the experiment with the squares, sticks, and cubes described above might do the following project on a balance scale and then write a report relating the two projects:

Balancing 7 weights on the seventh peg, 7 weights on the fourth peg, and 4 weights on the first peg (all on the left side) against 9 weights on the ninth peg on the right side [$x^2 + 4x + 4 = (x + 2)^2$]. Balancing 6 weights on the sixth peg, 6 weights on the fifth peg, and 6 weights on the first peg (all on the left side) against 9 weights on the eighth peg on the right [$x^2 + 5x + 6 = (x + 3)(x + 2)$].

4. *Recall.* Other activities should allow students to recall facts or to repeat the execution of steps in a process. Once a student is able to execute the steps in a process, he often gains confidence and self-esteem through successful involvement in drill activities related to the process. Students might work with flash cards or play a game in which numeral cards are drawn from a pile and placed in an addition or multiplication table. Naming shapes, taking memory tests, and working cross-number puzzles also reinforce knowledge. Other activities will help students pinpoint difficulties in using skills and provide practice to help remedy the difficulties.

5. *Application.* Some activities should involve the application of mathematical ideas. These would include the solving of hypothetical and real problems found in textbooks and books related to the history of mathematics, seeking answers to questions that are related to the

environment of the students themselves, and investigating real problems that are applications of the topic under study. There are countless fascinating examples of problems to which students can apply their knowledge of mathematics. These include the weight problem of Bachet de Méziriac (Dörrie, pp. 7–9); finding the amount of pressure a student exerts on the floor by standing on two feet, one foot, or the ball of one foot; finding the rate at which some specific prices are rising; and making clocks (or other objects) that can be used to clarify modular arithmetic (or other concepts developed in class).

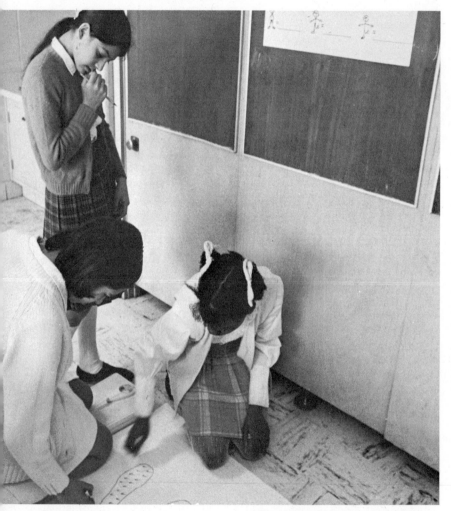

How much pressure do you exert on the floor when standing on both feet?

6. *Planning, Evaluation, and Remediation.* Other classroom activities involve planning, evaluation, and remediation. The teacher should have each student accept much of the responsibility for his own learning. Students should be given opportunities to share in class planning, and they should be allowed to make some choices. The teacher must establish class goals, and he must carefully guide the students toward their individual goals. Each student should be helped to evaluate his own efforts and achievements. The teacher should provide for diagnosis of difficulties and suggest ways of remedying them. He must also help each student evaluate himself and help him remedy his own difficulties.

A plan such as the following might be adopted by a mathematics teacher for apportioning class time for the various types of classroom activities. This is only a suggestion. The apportionment plan for any class must take into consideration many factors peculiar to that class: skills of the teacher, communication levels of the students, number of students, availability of facilities, behavior patterns of the students, and the level of mathematics being taught.

Percentage of Class Time Allotted to Various Types of Activities in the Mathematics Classroom

	TOTAL CLASS ACTIVITY	LABORATORY OR INDIVIDUAL ACTIVITY	
	teacher-student discussions, planning, concept development, integration of ideas	readiness, practice, diagnosis, remediation, reinforcement	problem solving, projects, extension of classroom ideas
More capable students	35%	20%	45%
Less capable students	25%	50%	25%

Fig. 1-10

Notice that a large part (65 to 75 percent) of the class time is allotted to individual or small-team activities. It is suggested that the less capable student devote about 50 percent of his time to readiness, practice, diagnosis, remediation, and reinforcement activities. The more capable student should spend about 45 percent of his class time applying mathematics concepts and skills and investigating ideas that might be considered extensions of the class discussions. The more capable student will spend less time than the less capable one on drill, readiness, diagnostic, and remedial exercises.

Small-team activities play an important part in the mathematics laboratory.

Rationale of the Laboratory Approach

The mathematics laboratory, with its emphasis on individual student manipulation of objects, brings a sense of realism into the mathematics classroom. Mathematics takes on a relevance and a significance for the student that are seldom present when the normal textbook teaching method is being used.

Through observing and handling physical objects, the student can gain experience from which he can construct mathematical concepts. Observing a student who is manipulating objects in order to find answers to questions is an effective way for a teacher to evaluate the student. The teacher can thus assess the student's skill in problem solving, his understanding of mathematical concepts, his skill in computing, his attitude, and his work habits.

In the laboratory the student can gain background in all aspects of problem solving. The surroundings of physical objects aid him in sizing up the problem and in selecting sound and feasible methods of attack. He also receives training in selecting data relevant to the question that has been posed and to his methods of attack. Furthermore, there often are built-in verifications that give him a feedback on the soundness of his method, the accuracy of his data, and the correctness of his computations.

By minimizing verbalism, which tends to bog down many students, the laboratory approach provides more opportunities for students to be successful than conventional classes. The laboratory activities can give status to the mathematics class. The mathematics laboratory is a suitable place for a wide variety of student interests. The student is given the opportunity to work at his own rate. He can use his own ingenuity and creativity to find answers to his questions. It is likely that activities in the mathematics laboratory will become the topics of conversation outside class.

2

Organizing and Supervising for the Laboratory Approach

It is not uncommon for teachers to fear or dislike mathematics. They make little effort to teach it well. Many other teachers consider it a simple matter to organize and teach a mathematics lesson. Both types of teachers use the textbook to delineate the scope and sequence of the course, outline the approach to teaching, and provide the major source of exercises and problems. These teachers then need only to explain the content of the day's lesson, assign practice exercises, and mark their students' work. Teachers who rely mainly on the textbook rarely see the need to make arrangements for the use of materials other than the textbook, chalkboard, and overhead projector. They see little need for students to move around in, much less outside, the classroom, or for students to discuss their work with one another. This type of teacher creates an environment in which he is recognized as the authority; he does the planning, makes the decisions, evaluates the students' efforts, and thereby assumes the responsibility for learning.

The organization and supervision of the class become much more complex if the laboratory approach is implemented. In a classroom where the laboratory approach is used, the students must assume a great deal of initiative for selecting a way to solve a problem. They must select materials and learn to manipulate them. They will also have to make observations relevant to the problem on which they are working. They must learn to remain orderly and businesslike while moving around the room getting out, working with, and putting away materials. They must learn to work in teams of two or three members, as well as in large discussion groups. They must also learn to discuss plans, observations, and conclusions with team members.

"You find things to measure, and
I'll finish the chart." Everyone has
a job to do for each activity.

Teachers could have difficulty making an immediate transition from
formal class methods to the laboratory approach. Unless their responsibilities have been outlined and they have been trained, many students
would have difficulty understanding and accepting the purpose of the
problem to be investigated. They would be confused and would not
know how to begin. Many would not know what materials to use or
where to find them. If they were not told how to attack the problem
or if the method of attack did not quickly become obvious, many students would give up. These would be the confused and immature, and
they would probably use their new freedom of movement to resort to
horseplay.

Teachers experienced with the laboratory approach report that
most students who were uninterested in mathematics in the conventional
classroom develop some enthusiasm for doing laboratory activities.
They caution that most difficulties in the laboratory result from students
not knowing what to do. This happens when the duties and responsibilities have not been clearly delineated, when students have not been
instructed in the use of equipment, when they don't know how to attack
problems, when several people want to use the same equipment at the
same time, and when materials are disorganized. This chapter should
help those wishing to implement the laboratory approach to avoid
these difficulties.

Delineating Responsibilities

The teacher must accept the responsibility for procuring materials (objects, charts, models, equipment, reference booklets), but he should use the resources available to him. For example, he might have the school secretary or a group of students help him identify materials and make out purchase orders. He must have the materials organized and stored in such a way that students can find them easily. Students, other teachers, paraprofessionals, and janitors can help set up and maintain storage areas. The teacher should tell students where the laboratory materials are, what the students' responsibilities are regarding these materials, and how their use is to be scheduled. He must teach students how to carry out laboratory investigations. He must show them how to schedule and use equipment. The teacher must evaluate attitudes, work habits, and accomplishments of each student (see chapter 5). He must prepare or procure guidesheets for the investigations (see chapter 3). Finally, he must supervise and guide the students.

Specific responsibilities should fall to the students. They should select materials for each laboratory activity and return materials to the proper storage area when finished. Each student must be responsible

Students enjoy selecting their materials and keeping their storage areas organized.

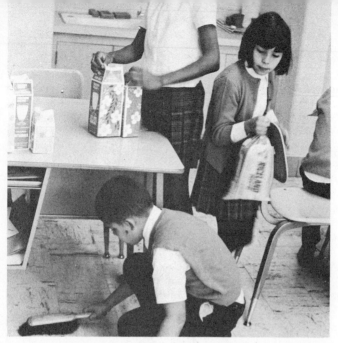

There are four cups in a quart, and
three people can clean one work
area in two minutes.

for keeping his own work area neat. Each must learn to work effectively
as a team member. He should keep a record of his work in an understandable and legible form; and he should evaluate his own work and
take responsibility for personal improvement in skills and understanding.
He should take it upon himself to develop an inquisitive attitude, and
to keep at a problem and not give up when encountering a block.

Instructing Students in the Laboratory Method

The laboratory approach is based on the premise that people need
to solve problems and that they can learn to do so. Problems require
knowledge of problem-solving techniques, and yet techniques vary from
problem to problem. Therefore it is difficult to train people to solve
problems. Any complex task (for example, piano playing or mountain
climbing) demands training, and problem solving is no exception. Training is provided by having the student solve problems (or play the piano
or climb mountains) under the direction of a teacher who is an expert
in solving mathematical problems (or playing the piano or climbing
mountains).

In the preface to his book on mathematical discovery, Polya has
written the following (vol. I, pp. v–vi):

Solving problems is a practical art, like swimming, or skiing, or
playing the piano: you can learn it only by imitation and practice. . . .

If you wish to learn swimming you have to go into the water, and if you wish to become a problem solver you have to solve problems.

If you wish to derive the most profit from your effort, look out for such features of the problem at hand as may be useful in handling the problems to come. A solution that you have obtained by your own effort or one that you have read or heard, but have followed with real interest and insight, may become a pattern for you, a model that you can imitate with advantage in solving similar problems. . . .

It may be easy to imitate the solution of a problem when solving a closely similar problem; such imitation may be more difficult or scarcely possible if the similarity is not so close. Yet there is a deep-seated human desire for more: for some device, free of limitations, that could solve all problems. This desire may remain obscure in many of us, but it becomes manifest in a few fairy tales and in the writings of a few philosophers. You may remember the tale about the magic word that opens all the doors. Descartes meditated upon a universal method for solving all problems, and Leibnitz very clearly formulated the idea of a perfect method. Yet the quest for a universal perfect method has no more succeeded than did the quest for the philosopher's stone which was supposed to change base metals into gold; there are great dreams that must remain dreams. Nevertheless, such unattainable ideals may influence people: nobody has attained the North Star, but many have found the right way by looking at it.

Identifying and stating the problem. First the student must be helped to identify problems. He must be encouraged to wonder about things, to raise questions, to seek explanations, and to make conjectures. For example, if the class wishes to play softball, how and where should they lay out the diamond so that the batter will not have to look into the sun? Tom's father jogs around the park. How can we measure the distance he runs if part of his trip is around a curved path? What are gear ratios? How does a computer work? Why do some people pay more for life insurance than others? How does an artist represent a three-dimensional object on a two-dimensional canvas? Why do astronauts use less fuel blasting off from the moon than from the earth?

Many times a student will state a question in very vague fashion. He must be helped to think about the nature of the answer he is seeking. He must be helped to state his question in simple, precise, and understandable form. It may be wise for him to write his statement of the problem before trying to find an answer.

In most cases the problem that the student is investigating in the laboratory will be stated completely or in part in the guidesheet. He should form the habit of thinking about the problem before he begins collecting data. He may be helped with questions like the following: What is the situation described here? What questions are raised by this problem? Is the problem like any others that I have solved? What will be the nature of the answers? What is my estimate of the answers? What data must I collect or what observations must I make in order to answer these questions?

You must think about a problem
carefully before you start to
collect data.

Finding a method of attack. Textbooks are usually arranged so that all problems in a section are to be solved by the same method. Therefore, students are given little encouragement to explore a variety of methods of attacking a problem; in fact, they may be led to believe that only one method exists. Investigations in the laboratory, as in real life, often utilize various methods. For this reason, laboratory activities help to eliminate the unrealistic one-method syndrome so characteristic of mathematics classes.

Students should be trained to think of a variety of ways of attacking a problem. They may propose ways that have never occurred to the teacher, and the teacher can encourage the brainstorming process by such comments as "That's a good idea, Ken," or "That's great, Shirley; I would never have thought of that." Each idea should be accepted and listed on the chalkboard. Then the class should evaluate the soundness and feasibility of each idea. Only after an idea has been discussed and each student has had his say should it be accepted or rejected. (Often the underachieving student may show surprising originality. His originality is what may have motivated him to be a low achiever in the first place. Many classrooms allow only *the* answer, each student must conform, and if a student's original suggestions are not accepted it's easier for him to stop participating than to go on making suggestions.)

Suppose, for example, the class has decided to attack the problem of finding the volume (in cubic inches) of cylinder A (fig. 2-1). The teacher might hold up one unit cube and say, "How many of these cubic inches would it take to construct a figure with the same volume as block A?" Some of the plans of attack that might be suggested are listed below.

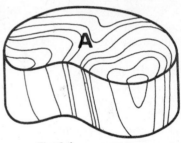

Fig. 2-1

1. *The weight methods.* Compare the weight of A with the weight of
   - a) 20 or 25 unit cubes, or
   - b) a larger block B of easily determined volume (for example, a rectangular parallelepiped).

2. *The layer methods.* Find the number of cubic inches in one 1-inch layer of A and multiply by the number of layers. To find the number in one layer, one may
   - a) place a layer of 1-inch cubes on top of A;
   - b) place a square-inch grid on A;
   - c) trace the outline of the top of A onto a piece of paper and measure the area within this outline; or
   - d) place a clear acetate sheet over the top of A and draw a rectangle that encloses a region that appears to have an area equal to that of the top of A.

3. *Displacement methods.* To find the volume of water displaced by A, one may
- *a)* submerge A in a full container, and measure the overflow water in a graduated cylinder or measuring cup; or
- *b)* submerge A in a large cylindrical or prismatic container and measure the change in water level.

Students should conclude that the first method is sound only if the blocks of wood are of the same density and if the unit cubes have been accurately cut. They may conclude that 1*b* is the most accurate of the weight methods.

The layer method will probably be the most meaningful to the students. Making the layer of 1-inch cubes should convince them of the efficacy of this method. Some may object that the method is not accurate and that it is not possible to measure the region enclosed in a figure such as the top of A. If this happens, a dialogue between students or teacher and students about measurement should bring about a healthy change in thinking, for most people do not understand measurement. Any reluctance of the students to measure area with a grid can be lessened by a similar dialogue, or by having them observe the agreement in the measurements obtained by repeated trials. A square inch drawn on a piece of clear acetate and subdivided into 16 smaller squares will enable the students to get more precise measurements of the area.

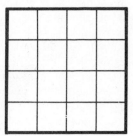

Fig. 2-2

The two displacement methods should appear sound if block A will not absorb water. Method 1*b* requires that students be able to calculate the volume of the displaced water.

Carrying through the plan of attack. Many students are disorganized, and inexperienced in planning and taking steps in an experiment. They also have trouble putting their results down in an understandable form. The teacher should guide them in planning as well as in obtaining and recording data. As examples, let us consider how we might want students to use methods 1*a*, 2*c*, and 3*a*, as stated in the preceding section.

Steps in using method 1a for an individual or a team:
1. Get postal scales that will measure in ounces to 5 pounds, and adjust the zero setting.
2. Find the weight of block A to the nearest $\frac{1}{2}$ ounce.
3. Find the total weight of 20 unit cubes to the nearest $\frac{1}{2}$ ounce.
4. Set up a proportion relating the number of ounces in each weight to the volume of each set of objects weighed.
5. Solve the proportion for the unknown term (number of cubic inches in block A).

Sample record of data:

Block A		20 unit cubes	
Weight:	__42__ ounces	Weight:	__8__ ounces
Volume:	☐ cubic inches	Volume:	__20__ cubic inches

(number of ounces : number of cubic inches)

(42 : ☐) = (8 : 20)

8 × ☐ = 42 × 20   (From the product rule)

$$☐ = \frac{42 \times 20}{8} \text{ cubic inches} = 105 \text{ cubic inches}$$

Fig. 2-3

Steps in using method 2c for an entire class:

1. Trace outline of top of block A onto a spirit master and run a copy for each pair of students.
2. Give each pair a square-inch grid drawn on clear acetate, a square inch subdivided into $\frac{1}{4}$-inch squares also on clear acetate, and a marking pencil.
3. Have each pair get the measure of the area within this region to the nearest $\frac{1}{4}$ inch squared.
4. Collect results. Classify the results and have those with results far above or below those of the modal class do their work over. Find the arithmetic mean of the measures in the modal class. This measure is also the number of cubic inches in a 1-inch layer.
5. Find the number of 1-inch layers.
6. Compute the volume of block A.

Sample record of data:

Measures of the area of the top of block A:

___ ___ ___ ___ ___ ___ ___

___ ___ ___ ___ ___ ___ ___

Make a histogram of the measures.

Take the arithmetic mean of the measures in the modal class.

_____ square inches

The number of cubic inches in a 1-inch layer is _____.

The number of layers is _____.

The approximate volume of block A is _____ cubic inches.

Fig. 2-4

How big is this block? Many problems can be solved in more than one way.

Steps in using 3a as a teacher demonstration for the whole class:
1. Shellac block A and the unit cubes (otherwise they will be spoiled for weight work).
2. Bring (or have a student bring) a bucket that is large enough to hold all of block A.
3. Bring (or have a student bring) a basin that is large enough to hold the bucket and catch the overflow.
4. Obtain a measuring cup or graduated cylinder.
5. Fill the bucket to the brim with water and place it in the basin.
6. Submerge block A by pushing it under with the eraser end of a pencil.
7. Pour the water from the basin into the measuring cup or graduated cylinder and record the measure.

Record of data (for a measuring cup calibrated in fluid ounces):

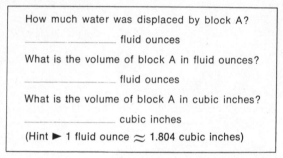

How much water was displaced by block A?

_____ fluid ounces

What is the volume of block A in fluid ounces?

_____ fluid ounces

What is the volume of block A in cubic inches?

_____ cubic inches

(Hint ▶ 1 fluid ounce $\approx$ 1.804 cubic inches)

Fig. 2-5

Follow-up: Drawing conclusions, comparing results, or comparing methods. After carrying through the three methods with the class, the teacher might ask the following questions: "What is the volume of block A? . . . Which of the three results do you have more faith in? . . . What could you do to get these results to be in closer agreement? . . . What is the importance of taking two or three readings of each measure? . . . Would the layer method work if block A were shaped like a rock? . . . Under what conditions would the weighing method be a sound one to use? . . . Would there ever be times when the displacement method was best, or when it would not work?"

Ways of Conducting a Laboratory Lesson

To have laboratory activities carried on with a minimum of confusion and a maximum of learning, the teacher must do much day-to-day planning. Planning will be discussed in connection with three different ways of conducting a laboratory lesson: teacher demonstration; all students in a class working in small teams on the same experiment; and students working in small teams on different experiments.

A teacher demonstration is not good unless students are interested and involved.

Teacher demonstrations. Teacher demonstrations are useful for teaching students how to formulate problems, analyze problems, select sound and feasible strategies for obtaining solutions, plan steps for reaching solutions, gather data, record data, and draw conclusions. The effectiveness of a teacher-directed demonstration depends on the students' interest in finding a solution to the problem, the suitability of the materials used, and the ability of the students to view the demonstration. It is important that the students be made to feel involved in the demonstration.

The teacher, first of all, must plan how to present a problem to the class. He must then select materials related to the problem. It is often necessary to use large-scale models of instruments (such as large graduated cylinders, vernier calipers, slide rules, and chalkboard instruments). The overhead projector can be used to present prepared drawings and tables, as well as the silhouettes of objects, to the entire class. Instruments should be accurate and in good working order. Any materials used in an experiment should be familiar to the students.

The teacher should plan the steps to be used in reaching a solution, but he should assign students to carry out many of the steps. For example, in using method 1*b* to find the volume of block A, one pair of students might weigh each block, another pair calculate the volumes (using ratios), and a third pair keep a record at the chalkboard.

The teacher must plan ways to get students involved. One technique is to ask them to recall relevant facts previously studied. For example, activity 1*b* might begin with questions about methods of finding the area of a rectangle, as well as the meaning of *cubic inch* and the number of layers of cubic inches in a solid. Students might be asked how to find the volume of a rectangular parallelepiped (block B). Having a student give an illustration of an idea may result in that idea's becoming more meaningful to him. Having a student offer a suggestion or making a conjecture or prediction may cause him to think about the kind of answer being sought and to feel some commitment to the activity. For example: "Suppose block A were sawed into cubic inches. How many cubic inches do you estimate there would be? . . . Would it be easier to cut block A or block B into inch cubes? . . . How would the weights of inch cubes taken from the two blocks compare?" How can a teacher get all students to react to the questions raised? Responses may of course be given orally, with the teacher being careful to call on each student. Some teachers might write key questions in an experiment on transparencies for the overhead projector and have the students give short answers on a response sheet. Students should be given an opportunity to check their responses as the class moves through the experiment.

The teacher must plan how he intends to organize and present observational data on the chalkboard or overhead projector. How should he label his data? Should he present data in tabular or graphic form?

Having a student apply some process or principle to a particular situation not only enables him to obtain an answer but also reinforces his understanding of that process or principle. For example, the ratio of the number of cubic inches to the number of ounces should be the same for blocks A and B. The use of equivalent ratios and the product rule should lead to the solution and at the same time offer an opportunity for the students to reinforce their understanding of equivalent ratios.

All class members manipulating identical sets of materials. When sufficient materials are available and when the teacher feels a need to give close direction to his students, the best choice may be to keep the entire class together. Teachers often use this method in exercises involving paper folding, geoboards, pendulums, slide rule construction, adjustable quadrilaterals, and fieldwork.

Plans must be made to obtain and distribute materials and to share equipment. It seems wise to have students work together in teams of two, three, or four, depending on the activity and the amount of available materials. Materials can be collected into kits and each team provided with a kit. The teacher must avoid delays at stations where such items as a stapler, glue, or a paper trimmer are located. This can be done by instructing students to work rapidly but accurately at these stations, by having the stations in different areas of the room, by having as many such stations as possible, and by having students at a busy station switch to a less busy one.

In general, this teaching strategy provides for student involvement under the close supervision of the teacher. General instructions, questions to be answered, and observations to be recorded can be provided orally by the teacher, as well as by the overhead projector, chalkboard, and hand-out sheets. As the class progresses, the teacher has the opportunity to interrupt in order to give additional instructions, discuss and evaluate student responses, allow students to ask questions, extend the activities, and collect and interpret observations.

Let us consider the planning that must be done by a teacher of a class of 25 eighth-grade students in introducing them to the use of the plane table and alidade. He must consider the following:

1. What type of fieldwork can be assigned to a class of this age and ability?
2. Where on the school campus may the class do its fieldwork?
3. What are the basic mathematical concepts and skills involved?
4. What must the students be taught beforehand?
5. How can one teacher supervise five teams of five members each when all members are inexperienced?
6. How should the materials be assembled?

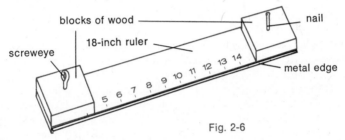

Fig. 2-6

This is an example of a simplified right-handed alidade. To make one—

1. Glue strips of $\frac{3''}{8}$ -thick wood to the nonbeveled portion at each end of an 18″ hardwood ruler with a metal edge.
2. Drive a $1\frac{1}{2}''$ finishing nail into the center of one block of wood.
3. Screw a $1\frac{1}{2}''$ screweye into the center of the other block of wood.

Fig. 2-7

The plane table is best supported by a tripod. If that is not available a three-legged stool or a tall wastebasket (as shown here) will do. A marble can be used as a level.

The teacher may decide to start with an angle-angle example of similar triangles in a simple "fire tower" problem. The problem will be to compute the distance from A to C when it is possible to see each point from both A and B and where the distance between A and B can easily be measured.

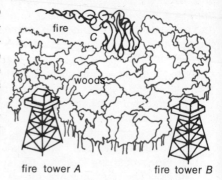

Fig. 2-8

The basic concept involved is that two triangles are similar if two pairs of corresponding angles are congruent. It follows that the ratios of the pairs of corresponding sides are equivalent. Worthy of mention is the idea of computing with approximate data. The more students are made aware of this, the better will be their concept of measure.

The problems can be set up on the football, soccer, softball, or baseball field. Half-sheets of notebook paper can be labeled for the corners of the triangle. Wooden stakes can be driven through the pieces of paper to locate the vertices of the triangle. Then the students would copy the triangle, using the plane table and alidade. (If a low support is used for the plane table the students should be supplied with a kneeling pad.) The stakes can be driven just before the class convenes and removed by the students at the end of the class period. Figure 2-9 shows five such problems set up on a football field.

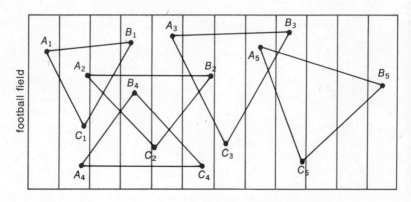

Fig. 2-9

Learning to use the alidade can be done in the classroom. A sheet of wrapping paper can be taped to the wall. Points *A*, *B*, and *C* can be located as the vertices of a triangle. The distance from *A* to *C* should be between 35 and 45 inches. The plan is to draw a triangle similar to triangle *ABC* on an $8\frac{1}{2} \times 11$-inch sheet of paper. To do this:

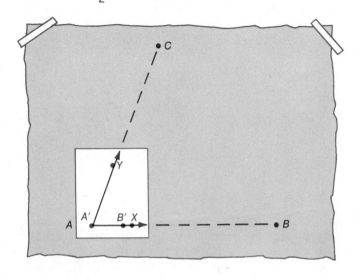

Fig. 2-10

1. Place the sheet of paper over *A* (see fig. 2-10). Mark point *A'* on the paper.
2. a) Stretch a string from *A'* toward *B* and line up *A* and *B*.
   b) Draw *A'X* so that *X* is in line with *B*.
3. Draw *A'Y* by sighting toward *C*. ∠*XA'Y* is a copy of ∠*BAC*.
4. Measure side *AB*. Select a scale to use and locate *B'* on *A'X*. Move the sheet of paper to *B* so that *B'* is over *B* and *A'B'* aligns with *AB*.
5. Repeat steps 1, 2, and 3 at *B* to construct angle *A'B'Z* (see fig. 2-11).
6. Mark the point at which *A'Y* and *B'Z* intersect. Label it *C'*.
7. Measure side *A'C'* and use a proportion to find the length of side *AC*.

Verify this measure by measuring segment *AC*. The computed value should be within one inch of the measured value.

Several skills must be taught prior to this activity. A pair of students might demonstrate how to use a 50-foot tape measure to measure a distance of 200 to 300 feet along a straight line. They must be familiar with the markings on the tape, especially the zero point. The person in the rear must keep the front person in a line with the stake toward

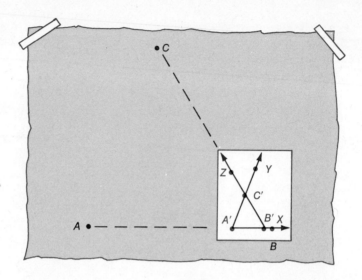

Fig. 2-11

which they are moving. Every 50 feet the front person leaves a marker, which may be a large nail punched through a colored piece of cardboard. Another pair of students might demonstrate how to use the alidade. The object is to draw two angles of a triangle on a sheet of paper so that the resulting triangle is similar to the original triangle. To do this—

1. Set the plane table on the support so a point on the paper is over point *A* on the ground. Mark point *A'* on the paper.
2. *a)* Align the alidade with the side, making sure the edge passes through *A'*.
   b) Draw *A'X* so that *X* is in line with *B*.
3. Draw *A'Y* by sighting toward *C* so that the edge passes through *A'*. ∠*XA'Y* is a copy of ∠*BAC*.
4. Measure the side *AB*. Select a scale and use it to locate *B'* on *A'X*. Move the plane table to *B* so that *B'* is over *B* and *A'B'* aligns with *AB*.
5. Repeat steps 1, 2, and 3 at *B* to construct angle *A'B'Z*.
6. Mark the point at which *A'Y* and *B'Z* intersect and label it *C'*.
7. Measure side *A'C'* and use a proportion to find the length of side *AC*.

Before sending the students out to do the fieldwork, emphasize the analogy between the stretching of the string and the use of the plane table and alidade.

Team captains should be appointed to assume leadership in the fieldwork. Each captain should be given a list of the materials needed by his team, a statement of the goals of the team's work, and a description of the jobs to be assigned to the team members.

Class needs and available materials
will determine whether students
all work on the same activity.

Students working as teams in separate activities. Students bring a variety of abilities, needs, and attainments to the laboratory. Therefore it is usually best to form small teams and assign them different activities. The composition and size of laboratory teams depend upon the type of activity, the materials available, and the manner in which the students work together. Some teachers make use of three-member teams grouped randomly. In scheduling activities, the teacher must take into consideration the logical sequence of the activities, the needs and preferences of team members, the number of sets of materials available, and the weather (if outside work is involved).

A large assignment chart such as the one that follows is helpful in scheduling activities. This chart refers to ratio activities for which all guidesheets are included in Appendix B. There are five basic level-A activities that must be completed before any of the five optional ones are done. The small numerals indicate the number of sets of manipulative materials available for the specific activity. A small check indicates that the team has spent one day on this assignment; the ⊠ indicates that the experiment has been completed. The small squares are movable pieces of red cardboard that represent the day's assignment; the triangle indicates a second assignment that can be done if the first assignment is completed quickly. The three calculators, mathematical games, and geoboards permit some flexibility in scheduling.

Class Assignment Chart

TEAM	STUDENT	CALCU-LATOR	BASIC R-A-1—R-A-5		R-A-6	OPTIONAL (2 or 3) R-A-7	R-A-8	R-A-9	R-A-10	MORE ADVANCED B LEVEL			GEO-BOARDS	GAMES
A	B. Adams	√		□										√
	T. Heston													
	B. Chaney													
B	H. Owens		√	√										
	J. Brown				□		△							
	H. Monroe													
C	B. Deal	√	√	□										
	P. Reed													
	D. Young													
D	R. Davis	□	√	√										
	D. Aaron													
	T. Landsman													
E														

RATIO

GAMES: *Numo*   triangular dominoes   3-D tic-tac-toe
       Ranko   game of multiples

Fig. 2-12

Rules for mathematical games should be kept with the games. Drill exercises and procedural flowcharts should be alongside each calculator. Guidesheets for the ratio activities should be in folders labeled and filed so that they can be easily located. Two bicycles (with three gears) are needed for experiments R-A-7, R-B-3, and R-B-12. Two bicycle wheels mounted on wooden forks should be supplied for experiments R-A-6, R-A-7, R-B-6, and R-B-12. Ratio compasses will be needed for experiment R-A-8. Meter sticks, foot rulers, 30-centimeter rulers, 50-foot tapes, and taping pins also should be supplied for these experiments. These could be kept in cardboard boxes labeled "Measuring Tools." An assortment of items (some in cigar boxes or large envelopes) for the various ratio experiments should also be included in the box.

If the teacher can anticipate what difficulties the students will have on any day, he might decide to develop the needed skills and definitions before these difficulties are encountered. It is also possible for a teacher to help his students learn the skills after they have recognized their deficiencies. Some of the deficiencies that will interfere with the ratio experiments can be alleviated if the experiments are preceded by activities related to linear measurement, decimal fractions, estimation of computational results, and the use of calculators.

The guidesheets for R-A-1 to R-A-5 are carefully programed. Students are instructed to manipulate materials and to record their observations and computations in the blanks provided. The optional ratio experiments include applications of equivalent ratios. Some of the guidesheets for these contain detailed suggestions; others merely present a problem to be solved.

As the laboratory session begins, the teacher should check to see which teams are ready. When all teams have materials and are at work, the teacher should feel free to move from group to group. He may engage the members of each team in dialogue to help them understand the problem and select a method of attack. He can then instruct them in the use of equipment and evaluate their understanding of the goals of the activity. He may soon be able to judge which students require immediate guidance when they are stumped and which ones are able to work on their own for some time trying to solve a problem.

What is learned by the teacher when observing the teams at work will be invaluable in the follow-up class discussions. The teacher may want to bring the class together after one or two days on the basic ratio experiments to discuss the key ideas involved. He may review the meaning of ratio and have students give some illustrations of ratios. He could then have the whole class work through a problem involving equivalent ratios.

The problem, for example, might be to use the bicycle wheel mounted on a handle to measure the distance between two stakes on the campus. A number of turns of the bicycle wheel can be used for a calibration run. The ratio of the number of turns the bicycle wheel makes to the number of feet of the run can be used to find the distance

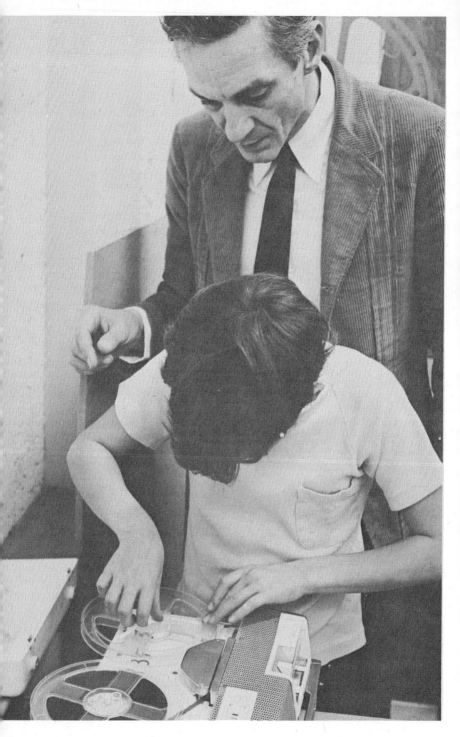

Measure your school with a bicycle wheel? You can if you know something about ratio.

between the stakes. This ratio can be set equal to the ratio of the number of turns of the wheel to the distance of the distance run (using the unknown □ for the distance). The experiment was performed in a ninth-grade general mathematics class working with the following data:

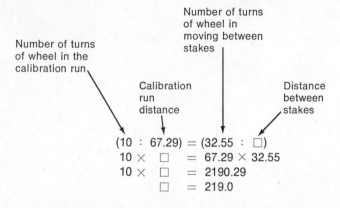

$$(10 : 67.29) = (32.55 : \square)$$
$$10 \times \square = 67.29 \times 32.55$$
$$10 \times \square = 2190.29$$
$$\square = 219.0$$

The distance between the stakes was approximately 219 feet.

Each student should keep a day-by-day log of his activities and a record of his laboratory work. If the guidesheets are punched, each student can keep them in his notebook. The main points of the class discussions can also be duplicated by the teacher and given to the students for their notebooks.

Students, especially low achievers, have difficulty in working toward long-range goals and integrating mathematical ideas. An effort should be made to help each student set a *daily* goal and work toward it. The teacher must also provide each student with frequent evaluations of his progress in the laboratory activities, although this must be done without forcing him to operate under a fear of making errors. Each student should be evaluated in terms of his willingness to benefit from his errors, the effort he makes to understand what he is doing (rather than merely filling in blanks), his willingness to take responsibility for improving his competencies, his desire for accuracy and thoroughness, and his diligence in working with his team, as well as his understanding of mathematical ideas. It may be desirable to make out a checklist to be marked in conference with the teacher.

Laboratory investigations may help some students identify their weaknesses—for example, inadequate skills with decimal fractions, inability to estimate, or poor understanding of angular measure. Skill reviews should be built into the laboratory situation so that they can be provided when the need for them is felt. Experiments should stimulate questions that students can answer by consulting the textbook or reference books, or by performing more experiments. Activities may suggest other topics for investigation. The teacher should be aware of interests aroused by activities. Learning is much more meaningful

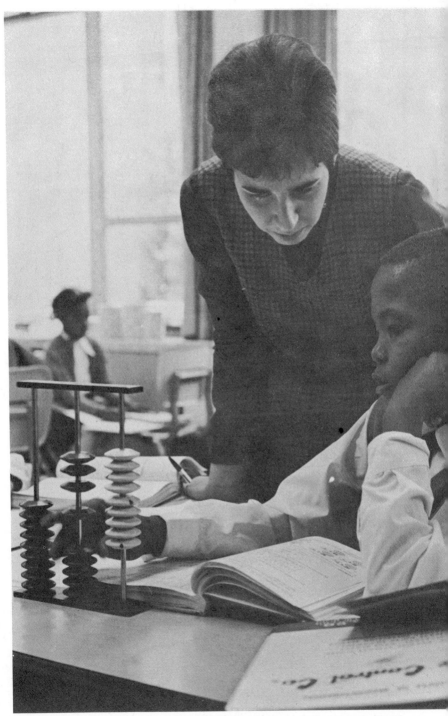

An abacus can help you understand
what happens when you add.

when a student's interest is piqued. For example, a student might become very interested in the history of linear measure or new methods and standards of measurement after an experience with guidesheets on linear measure. The teacher should provide references and other resources (such as information on visiting a nearby scientific laboratory) for such a student. Another student might become very interested in statistics, and similar resources should be provided for him. But this is all part of the evolving nature of classroom management. This is the most difficult part of organization and supervision because of its spontaneity, but it can also be the most productive and satisfying.

3

Planning Laboratory Investigations

The preceding chapters described the laboratory approach, how it fits into the curriculum, and how a teacher can organize and supervise a class for this approach. Much remains to be discussed: What changes in the classroom are necessary? What supplies are needed? How are methods of evaluation and grading affected? What does the approach do for low achievers? The remaining chapters deal with these questions. First, the steps involved in planning a laboratory investigation will be considered.

There are many different types of investigations and many sources from which a teacher can get ideas for investigations. Before deciding on an activity, the teacher must determine the criteria for the selection of a problem to investigate. He must then determine the best means of presenting the problem to the students. He must also consider how he can best help students make discoveries, draw conclusions, and report on their results.

Selecting Ideas for Investigation

Where does the teacher get ideas for an investigation? Perhaps the best place to look first is to the students' interests. Casual conversations with students will reveal topics of interest to them that can easily be utilized in a mathematics investigation. For example, most students are fascinated by games of chance, which can be the basis for the study of probability. Coins, decks of cards, and dice can be used to generate probabilities. (In cases where school policy outlaws the use of cards and dice, objects such as tacks, marbles, toothpaste caps, corks, and spinners can be used.) Such a study provides practice in computing with fractions and teaches probability distributions (and the futility of betting against the house). Other students might wonder about locating hidden treasures, designing stage scenery, or finding the distance across a lake. Such investigations can generate many topics related to geometry, such as locus in a plane, locus in space, perpendiculars, angle measure, and distance.

The teacher's creativity is a source of ideas for investigation. While investigating a new piece of equipment like a geoboard, the teacher might decide to have students use the geoboard to investigate many topics in geometry—for example, convex and concave polygons; interior, exterior, and boundary of a polygonal region; area of a polygon (drawn on a lattice point grid) as a function of the number of boundary points and interior points; Euler's theorem for polyhedra (as applied to polygons); and graphing in a coordinate plane. While using maps to determine the best route for a trip, the teacher might decide to use maps and globes as sources of problems regarding distance and scale. He might have the students investigate the use of little wheels or string to measure distance on a map or globe. He might also have them prepare scale drawings using ratios.

Discussions with other teachers can be another valuable source of ideas. Conversations about books of puzzles, problems, and games might suggest activities that would help students attain certain objectives; for example, there are games involving the grouping of objects of the same shape from a set of objects of different shapes and sizes, which can be played by students in the upper elementary grades. The objects can be both two- and three-dimensional. This activity is a good preliminary to the study of similarity and, consequently, to the activity with maps and globes mentioned above. Also, teachers might find ideas for activities that they think would be of particular interest to the students and that the students are capable of handling; for instance, junior high students usually enjoy working with shapes, reflections, rotations, symmetry, and the groups generated by objects of the given shapes as they are rotated and reflected. If the ideas help to develop learnings the teacher considers to be valuable, then these might be used even if they do not fit directly into the curriculum at that point in time.

Teacher discussions of this sort are also valuable for generating new ideas. One teacher might mention that geometry activities are needed. Another might add that one of the focuses should be mappings of points, and another might suggest a guidesheet on perspective drawing. All three would have contributed to the development of an investigation. The same teachers might come up with several more ideas for investigations in geometry appropriate to the elementary school, such as activities involving tangrams, polyominoes, soma cubes, the divine proportion as developed from the Fibonacci sequence, and curve stitching.

Activities must meet certain criteria. They should be interesting for the student and at the same time should be related to the instructional goals of the course. They must be at the right level of difficulty, and they must be practical, taking into consideration the available time and equipment.

Relation to the goals of instruction. In planning laboratory activities, the teacher must first state the goals of the course in precise terms; a careful study of the textbook or course syllabus should enable him to generate such a list. Some of these goals will be prerequisites to others and so must be attained first.

A laboratory activity may contribute in several ways toward attainment of the course goals:

1. The experiences may provide background for the later development of concepts. Such activities are relatively unstructured and often include games.
2. Many experiences are selected so that the student can make discoveries and generalizations. These activities are organized to lead the student directly to the discovery of some mathematical concept.
3. Other experiences require a student to apply some concept or process that he has learned. Through applications, mathematical ideas become more meaningful and significant, and the student's understandings and skills are extended and reinforced.
4. Laboratory experiences can be used to challenge a student, to provide him with opportunities for developing self-confidence and habits of independent work and for enjoying mathematics.

Certain topics should be developed at the beginning of the school year because they are valuable tools for the study of other topics. Ratio, as a method of comparing two sets of objects, is very useful, but most students do not understand it. The concept of mathematical sentences should be developed early. Common geometric shapes and their formulas also should be studied early in the year. Tabulation of data and some elementary concepts of statistics (such as mean, median, mode, range, and graphs), as well as measurement and approximation, should be considered. Another necessary skill that should be developed as early as possible is graphing in a rectangular coordinate system.

A series of guidesheets on ratio is included in Appendix B. These guidesheets should provide a basis for the development of this topic in the upper elementary grades. Teachers wishing to implement the laboratory approach might use these sheets to help them get started. (The work with volume described in chapter 2 requires a knowledge of ratio.)

Mathematical sentences are important because they provide a language for the analysis of data. Using equations, the student can formulate his data in an economical way. Using inequalities, he can concisely indicate the domain of a variable. The earlier he is presented with these ideas, the earlier he will have them to work with.

Playing with colored sticks can provide background for addition and multiplication.

Straight pins and spools of thread can help students make discoveries about geometry.

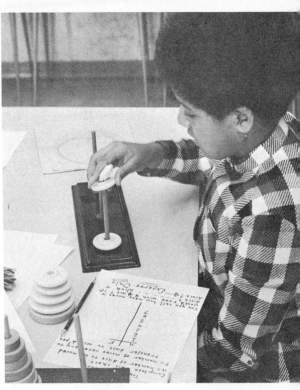

How many moves will it take to transfer eight disks? Mathematics can be used to make predictions.

How many beans? Simple questions can provide big challenges.

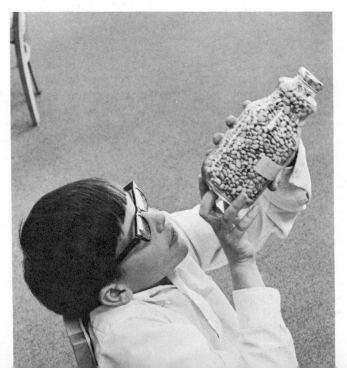

Shapes—drawing them and cutting
them out will help you remember
what they look like.

Much of elementary geometry concerns simple geometric figures
—recognition, knowledge of their properties, and generalizations about
them. Examples of activities that can be used to bring about these
learnings are work on the geoboard (including area); perspective
drawing; construction of polygons using sticks; counting parts of
polyhedra to develop Euler's formula; work with geometric solids (including volume); work with attribute blocks (including similarity); work
with cutouts of regions involving rigid motions; measure using the
alidade; measure using the trundle wheel; curve stitching; and tangrams.

Since most laboratory work calls for the collection and interpretation of data, students must know the rudiments of statistics. They
must be able to record and report data; they must be able to find the
mean, the median, the modal classes, and the range of data. Some of
the volume work in the preceding chapter required some knowledge
of statistics. Students should understand the relation between statistics
and measure.

The first laboratory unit might include the construction and use of
a ruler. If the ruler is marked in decimal units and a calculator is
available, tedious computation with fractions can be avoided. In the

Making a ruler can help you learn
about linear measure. It will be
used often.

construction and use of his ruler, the student can also gain some
understanding of decimal fractions (by making the divisions tenths).
An activity such as this can be extended to two- and three-dimensional
measure by construction of transparent grids and unit cubes.

Class sessions should periodically be devoted to approximate calculations. For example, 3.2×5.87 could be approximated by 3×6;
$\frac{18.42 \times 3.59}{8.63}$ by $\frac{18 \times 4}{9}$; $.875 \times 33.142$ by 1×33; and the cost of 14.6
gallons of gasoline at $.339 per gallon by $15 \times \$.34$. Work with relative
error would be meaningful after such a sequence of problems. For
example, the relative errors in the preceding calculations could be
found. This would be both meaningful and enlightening, since it can
show how large an error can get even though the change that caused
it is small.

Finally, the students should be taught graphing in a rectangular
coordinate system, which will allow them to see patterns pictorially.
For example, using measurement, the tabulation of data, and graphing,
they can derive an approximation for π. (See the guidesheet on functions in this chapter.) Similarly, junior high school students would

have difficulty seeing the relation between the period of a pendulum (T) and its length (l); but they can graph T^2 against l on the same sheet as the original graph and come out with an expression like $T^2 = Kl$ (where K is a constant), from which they can derive $T = c\sqrt{l}$ (where $c = \dfrac{2\pi}{\sqrt{g}}$ and g is the acceleration due to gravity in feet per second squared).

The teacher should be sure that mathematical ideas are studied in a logical sequence, so that competencies needed in a series of investigations will be available when they are needed.

Proper level of difficulty.   The teacher must be aware of the prerequisites for an investigation, and he must know enough about each student to decide whether he has sufficient competency to make the investigation worthwhile. Involvement in an investigation may show the student certain deficiencies and motivate him toward improvement. However, if there is too great a deficiency in his understanding of mathematical concepts, he cannot proceed effectively.

Interest to students.   Activities should be chosen in view of the appeal they have for the age group involved. Generally students in grades 5 through 9 prefer investigations that involve movement and action. Students in these grades are very curious. They like competition as long as they know that they have a chance to win. Also of particular appeal are experiments with built-in verification: How far is it from point A to point B? Which of these three dice are loaded? Can you play this game so that you will always win? Of what two-digit number am I thinking?

The teacher may take advantage of current topics of interest to start the class thinking about investigations. Space explorations: How large is the moon? How does one compute orbital speeds of satellites? What are escape velocities, and why is there a difference between the escape velocities on the earth and on the moon? Athletics: I wonder who in the class can throw a baseball the fastest. How much momentum does a 250-pound tackle have when traveling 15 miles per hour? Science: How much invisible water vapor is there in our classroom? If the classroom were sealed, how long would the oxygen last?

Relation to available tools.   The teacher or student can raise many questions for which he has neither the time nor the equipment, materials, or data necessary for the investigation. Before an investigation gets under way, questions such as the following should be raised: How can we find an answer? What kinds of materials are needed? Can we obtain these materials? Do we know how to use the equipment, or can we find out how? Where can we find the necessary data?

Time limitations must also be taken into consideration. There are valuable activities that will take more time than is available or more than the teacher had planned to spend on the topic. If a different activity would achieve the same results in significantly less time, it should probably be used instead.

How high does a ball bounce? Some activities involve action.

If there's a good chance to win, competition can make an activity more fun.

How many Beans

Preparing Guidesheets for Student Investigation

Some investigations grow out of questions raised in classroom discussions. When this happens, it seems advisable that the problem for investigation be carefully stated and written on the chalkboard. Most investigations, however, probably should be planned ahead of time by the teacher. This planning may be done in accordance with a statement of the goals of instruction of a topic of study.

Suppose, for example, a sixth-grade class is studying linear measure. The objectives can be listed in behavioral or conceptual terms. Behavioral objectives for laboratory investigations are discussed in chapter 5 and Appendix B. Below is an example of a statement of objectives in conceptual terms for linear measure.

Objectives in the Study of Linear Measure

1. Line segments have length.
2. Measuring a line segment requires the following two steps:
   a) Select a line segment of arbitrary length as a unit of measurement.
   b) Find the number of these units that would fit end to end along the segment to be measured. This number is the measure of the line segment.
3. All measurements are approximate; there will usually be part of a unit left over in the measurement process.
4. To ensure that the leftover part is small, a small unit of measurement must be used. *Precision* is a term used to refer to the size of the unit of measurement. The smaller the unit, the more precise the measurement.
5. Units of measurement are often divided into fractional parts to obtain greater precision.
6. To compare the lengths of two line segments, compare their measures.
7. The measure of a segment varies inversely with the length of the unit of measurement used.
8. A ruler is an object whose edge represents a line segment on which units have been marked off. The marked edge has the following characteristics:
   a) There is a zero point.
   b) It is labeled with numerals to indicate the number of units from the zero point to the points at unit intervals on the line segment.
   c) The units are sometimes divided into fractional parts for greater precision.
9. Some standard measuring devices are a 12-inch ruler, a meter stick, and a 50-foot tape.
10. There are many other devices for measuring length—for example, measuring wheels (useful for measuring along curved paths as well as for measuring longer distances) and vernier calipers.

11. Estimating length (or height or distance) is a matter of making a good guess of the number of standard units contained in the measure of a given segment.

12. Man's efforts to develop and standardize units of linear measurement have been both ingenious and comical.

The teacher may then plan activities for the students that will help them attain these objectives. If students are to work in teams on different investigations, it becomes necessary to develop some way of giving suggestions. Written instructions are given on *guidesheets,* and (if the school has tape recorders) verbal instructions can be given on tape. The guidesheets should give a clear statement of the objective to be investigated, a brief listing of the materials needed, and some indication of the procedure to be followed. Beyond this comes the difficult question of deciding how detailed should be the instructions on collecting, organizing, and interpreting the data. It seems advisable to direct the thinking of students if the object of the investigation is to lead to a generalization or develop a concept. Students can be given more of a free hand when they are exploring problems of special interest to them or applying what they have learned.

Guidesheets *must* be readable; they must communicate. All questions, instructions, and suggestions should be as brief as possible. Pictures should be used whenever possible.

The guidesheets should appeal to the students. Their reactions should be: This looks interesting. I understand the question. I wonder what the answer is? I should be able to find out. To elicit reactions such as these, the guidesheet should appear uncomplicated, it should present a challenging question, and it may contain color and sketches. (Color can easily be provided by using spirit masters of different colors to do different sections of the guidesheet.)

A guidesheet should promote active thought. Students can be asked not only to make observations but also to make predictions or conjectures to be checked. They can be asked to comment on results, form judgments, observe patterns, make comparisons, and draw conclusions.

Seven examples of guidesheets and the accompanying sections of teacher's manuals follow.

LM–1   *Linear Measure*

problem: How do you make and use a cool ruler?

materials:

1 sheet of ruled notebook paper
1 ice-cream stick
1 roll of adding machine tape
2 sticks, one green and one yellow

procedure:

1. Starting about 2 ice-cream stick lengths from one end of the adding machine tape, mark off 10 or 11 ice-cream stick lengths along the adding machine tape. Label the marks as shown in figure 1.

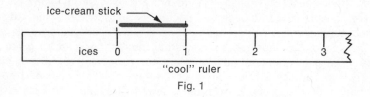

Fig. 1

2. On a small piece of paper mark off one "ice" (ice-cream stick length), and use the sheet of notebook paper (as shown in fig. 2) to divide the ice into tenths. Use the divided ice to divide the units on your ruler.

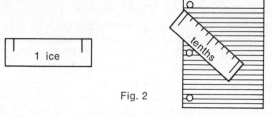

Fig. 2

3. One student's ruler is shown in figure 3; another's is shown in figure 4. The first student used his ruler to measure a stick, as in figure 5. The second used his to measure the same stick (figure 6).

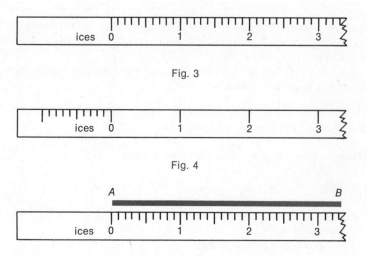

Fig. 3

Fig. 4

Fig. 5

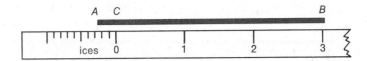

Fig. 6

In figure 6, how long is part *CB*? _____ ices
How long is part *AC*? _____ ices
In figure 6, how long is stick *AB*? _____ ices
In figure 5, how long is stick *AB*? _____ ices
Are the two measures the same? _____

    4. Make a ruler like that of the first student. Then make one like that of the second student.
    5. Why would anyone prefer a ruler like that in figure 3?

_____

    6. Why would anyone prefer a ruler like that in figure 4?

_____

    7. Measure the green stick. _____ ices
       Measure the yellow stick. _____ ices
       The green stick is about _____ times as long as the yellow
one.
    8. Estimate your height. _____ ices
       Measure your height. _____ ices
    9. Estimate the length of the chalkboard. _____ ices
       Measure the length of the chalkboard. _____ ices
   10. Measure the height of the door. _____ ices
       The length of the chalkboard is about _____ times the
height of the door.

F–1   *Functions: Parts of a Circle*

problem: How do you draw a circle whose circumference is 100 cm.?

materials:

    2 pieces of poster board taped end to end
    10 sheets of cm. squared graph paper (8½″ × 11″)
    Meter stick
    Felt-tip pen
    Colored yarn
    Scissors
    4 or 5 circular objects with different diameters
    String
    Cellophane tape
    Compass with extension arm
    Map distance wheel marked in centimeters

procedure:

1. Construct a large chart like that in figure 1. Trim the borders of the graph paper to fit.
2. *a)* Place each circular object on the chart. Draw around the object so that the circle passes through 0 and its center is on the base line of the chart. (See fig. 2.)
   b) Wrap a piece of yarn around each circular object. Cut each piece of yarn so that it fits around the object without stretching.
   c) Tape each piece of yarn onto the chart as shown in figure 2.

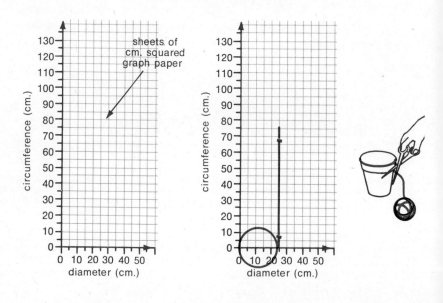

Fig. 1   Fig. 2

3. Use the compass to draw two or three circles as described in *a, b, c, d,* and *e* below.
   a) The measures of the diameters must be less than 50 cm.
   b) The circles must all be different sizes.
   c) The measures of the diameters must be different from the measures of the diameters of the objects in step 2.
   d) Each circle must go through 0.
   e) The center of each circle must be on the base line of the chart.
4. *a)* Use the map distance wheel to measure the distance around (circumference of) each circle you drew with the compass.
   b) Using the meter stick and scissors, cut pieces of yarn equal in length to each circumference.

c) Tape the pieces of yarn to the chart as you did in step 2.

5. You should now have six to eight pieces of yarn taped to the chart. Each piece should begin at the base line and extend upward. What do you notice about the top ends of the pieces of yarn?

6. Cut a piece of yarn 100 cm. long. Place it on the chart. If you had a circle whose circumference was 100 cm., the measure of its diameter would be about _____ cm. The measure of its radius would be _____ cm.

7. Draw a circle with this radius. Its circumference should have a measure of _____ cm. Measure the circumference. _____ cm.

F–1a   *Functions: A Measuring Wheel*

problem: How do you make a wheel for measuring short distances in meters?

materials:

The materials needed for guidesheet F-1, Functions: Parts of a Circle

The chart that resulted from doing F-1, Functions: Parts of a Circle

$\frac{1}{4}$″ plywood (15″ × 17″)

Coping saw $\frac{1}{4}$″-bolt, 2″ long
Sandpaper   2 washers
Masking tape
Handsaw   Piece of soft wood ($\frac{1}{2}$″ × 1″ × 14″)

Hand drill with $\frac{5}{16}$″ bit $6\frac{3}{4}$″ wire brads

Hammer   Large sheet of poster board (24″ × 38″)

procedure:

1. Using the chart constructed for F-1, Functions: Parts of a Circle:
    a) Cut a piece of yarn that is 1 meter (100 cm.) long.
    b) Find where the yarn would be on the chart if it represented the circumference of a circle.
    c) Tape it on the chart in the correct position and write the diameter of the circle (to the nearest cm.) to which it would correspond.

2. Use the compass to construct a circle whose circumference is 1 meter (100 cm.).

3. Using the map distance wheel, measure the distance around the circle you have drawn. Does it measure 1 meter? If it is more than $\frac{1}{2}$ cm. off, repeat steps 1, 2, and 3.

4. Being very careful, copy the circle on the sheet of plywood (see fig. 1). Mark the center of the circle on the plywood.

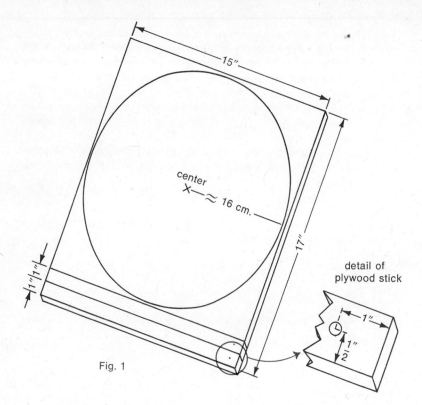

Fig. 1

detail of
plywood stick

5. Draw two sticks (each 1″ × 15″) on the sheet of plywood (see fig. 1).

6. Using the coping saw, cut out the circle. Use the sandpaper to make the circle smooth.

7. Measure the plywood circle, using the map wheel. If its circumference is less than 1 meter, wind masking tape around it. If it is greater than 1 meter, keep sanding it down until it is accurate.

8. Using the handsaw, cut out the two sticks.

9. Drill a hole in the center of the plywood circle.

10. Drill holes in the plywood sticks (see fig. 1).

11. Assemble all the parts as shown in figure 2.

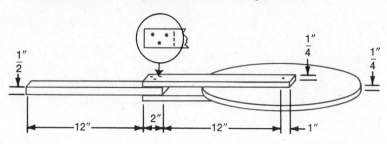

Fig. 2

12. Using the distance wheel:
   a) Measure the distance from where you are working to the wall on your right.
   b) Measure the distance from where you are working to the wall in front of you.
   c) Measure the sides of the classroom.
   d) Draw an outline of the classroom on a sheet of poster board, using a scale of 1 inch to 1 meter.
   e) Using the same scale and the measures from *a* and *b*, place a dot in the room outline that indicates where you are working. Have your classmates do *a*, *b*, *c*, and *d*.

P–1   *Pentominoes*

problem: How many different pentominoes can you make?

materials:

Masking tape
60 one-inch squares
Mirror
1 one-inch cube

procedure:

1. A figure that fits the following conditions is a pentomino.
   a) It is a single piece.
   b) It is two-dimensional.
   c) It is constructed from five squares.
   d) Each of the five squares has an edge in common with at least one of the others.
   e) Squares that touch have either an edge or a vertex in common.

Fig. 1

   The object in figure 1 is a pentomino.

Are the objects in figure 2 pentominoes? Explain why or why not for each.

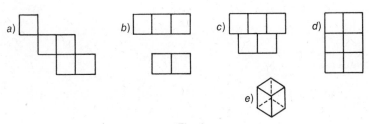

Fig. 2

2. Using your squares, arrange as many different pentominoes as possible. When you are sure an arrangement is a pentomino, and that it is not the same shape as any of your other pentominoes, tape it together.

3. Stand the mirror beside each pentomino (see fig. 3). Is the mirror image a different pentomino?

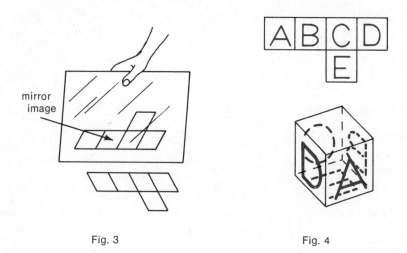

mirror image

Fig. 3   Fig. 4

4. Which of the pentominoes that you constructed can you fold around the one-inch cube? The cube has six faces. The pentomino must cover five of these faces (see fig. 4).

P–2   *Shapes from Pentominoes*

problem: What shapes can be formed with a set of pentominoes?

materials:

The set of 12 pentominoes you constructed for guidesheet P-1, Pentominoes

$8\frac{1}{2}'' \times 20''$ sheet of paper ruled into one-inch squares

procedure:

1. On your sheet of squares draw a rectangle 5 inches by 6 inches. Can you cover the 30 squares inside the rectangle with 6 of your 12 pentominoes?

2. Draw a 3 × 20 rectangle on your sheet of square inches. Can you use all 12 pentominoes to cover the rectangular region?

3. Draw a 4 × 15 rectangle on your sheet of square inches. Can you use all 12 pentominoes to cover the rectangular region?

4. Draw a 5 × 12 rectangle on your sheet of square inches. Can you use all 12 pentominoes to cover the rectangular region?

5. Draw a 6 × 10 rectangle on your sheet of square inches. Can you use all 12 pentominoes to cover the rectangular region?

6. Draw an 8 × 8 square on your sheet of square inches. Can you use all 12 pentominoes to cover the square region, leaving any 4 uncovered?

7. Can you repeat exercise 6 leaving a 2 × 2 square region uncovered in the center of the 8 × 8 square region?

AC–1   *Attribute Cards*

problem: How are the cards in the envelope alike?

materials:

A set of attribute cards (32 cards of different colors, shapes, and sizes, plus 3 charts)

procedure:

1. Place the 32 cards in the four spaces on chart I. In each space, all the cards must be alike. The cards in each space must be different from those in the other spaces.

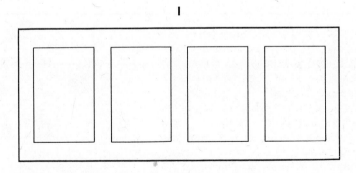

Fig. 1

2. Can two students place the cards in chart I differently and both be right? Explain. _____
_____

3. Each card has color, shape, and size. There are _____ different colors, _____ different shapes, and _____ different sizes.

4. Place the large red triangle in column A of chart II. In column *a* place all of the cards that differ from the large red triangle only in color. Must the cards in column *a* be large? _____ Must they be triangles? _____ Can they be red? _____

II

A	a	b	c	d	e	f	g
	color only	shape only	size only	color and shape	color and size	shape and size	color, shape, & size

Fig. 2

5. In column *b* place the cards that differ from the large red triangle only in shape.

6. Complete chart II. How many cards did you place in each column?

_____   _____   _____   _____   _____   _____   _____
   a   b   c   d   e   f   g

7. Chart III is a large square divided into 16 smaller squares. Each small square is numbered. Name (by number) the small squares that have a side in common with square 7. _____ Cards in squares that have a common side differ in only one way. The card in square 3 differs from the card in square 7 in size only. The card in square 8 differs from the card in square 7 in color only. Set up chart III like the one in figure 3. Place cards in the remaining squares. The card in each square must differ in one way from the card in each square with which it shares a side.

III

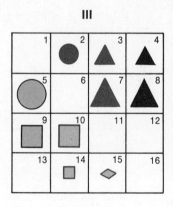

Fig. 3

III

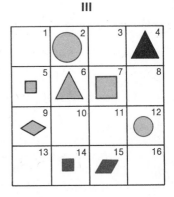

8. Set up chart III as shown in figure 4. The cards in squares that share a side must differ in exactly two ways. (For example, the card in square 2 differs from the card in square 6 in color and _____.) Place cards in the remaining squares.

Fig. 4

Games with attribute blocks are challenging to students and can become very competitive.

G–1   *The Geoboard*

problem: How do you find the area of a polygon on the geoboard?

materials:

    Geoboard
    Rubber bands

procedure:

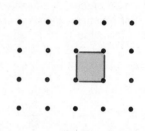

Fig. 1

    1. Stretch rubber bands around some of the pegs to form different shapes. If the rubber band does not cross itself, the figure is a polygon. The sides of the polygon are line segments.

    2. Stretch a rubber band around four pegs as shown in figure 1. The area of this polygon is one square unit. Make a polygon that has an area of

      a) 2 square units.
      b) 5 square units.
      c) $1\frac{1}{2}$ square units.

    3. Refer to figure 2. Duplicate each polygon on your geoboard. Which polygons have an area of $2\frac{1}{2}$ square units?

    a) _____   *b)* _____   *c)* _____   *d)* _____   *e)* _____

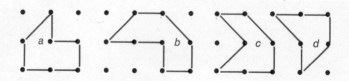

Fig. 2

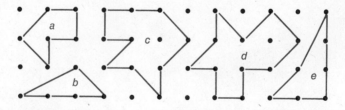

Fig. 3

4. Refer to figure 3. Duplicate each polygon on your geoboard. Find the area of each polygon.

a) _____ b) _____ c) _____ d) _____ e) _____

5. If you form a polygon on the geoboard that has no pegs inside except those touching the rubber band, can you predict its area? (The rubber band and the pegs that it touches form the *boundary* of the polygon.)

a) Construct a polygon with 4 pegs on its boundary whose area is 1 square unit. Construct other polygons with 4 pegs on their boundaries. What are the areas of the polygons?

b) Construct several polygons each with 5 pegs on its boundary. What is the area of each polygon? Fill in table I, where △ means number of pegs in the boundary and ☐ means area (in square units).

c) If two students each make a polygon that has 8 pegs on its boundary, must the areas of the two polygons be the same? _____

d) If you construct a polygon with 15 pegs on its boundary, the area will be _____ square units.

6. Refer to table I. Which number sentence expresses the relation between △ and ☐?

a) $\triangle = 2 + \square$ b) $\frac{1}{2}\triangle - 2 = \square$ c) $\frac{1}{2}\triangle - 1 = \square$

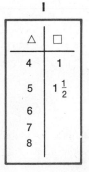

Fig. 4

△	☐
4	1
5	$1\frac{1}{2}$
6	
7	
8	

I

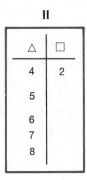

Fig. 5

△	☐
4	2
5	
6	
7	
8	

II

7. Complete table II. This time each polygon must have exactly one peg within it that is not touched by the rubber band. $\triangle$ represents the number of pegs on the boundary, and $\square$ represents the area of the polygon (in square units).

a) Construct a polygon with 4 pegs on its boundary whose area is 2 square units.

b) Construct a polygon with 5 pegs on its boundary whose area is $2\frac{1}{2}$ square units.

c) Construct a polygon that has 8 pegs on its boundary.

d) Predict the area of a polygon with 15 pegs on its boundary. _____ square units.

8. Refer to table II. Which number sentence expresses the relation between $\triangle$ (number of pegs on the boundary) and $\square$ (area)?

a) $\frac{1}{2}\triangle = \square$ b) $\frac{1}{2}\triangle - 2 + 1 = \square$

9. Form polygons that have exactly 2 pegs within them that are not touched by rubber bands. Enter your results in table III. Refer to table III. Write a number sentence using $\triangle$ and $\square$ that can help you find the area ($\square$) of a polygon when $\triangle$ represents the number of pegs on the boundary.

FOR EXPERTS ONLY. Write a number sentence using $\triangle$, $\bigcirc$, and $\square$ that can help you find the area ($\square$) of a polygon when $\triangle$ represents the number of pegs on the boundary and $\bigcirc$ represents the number of pegs within the polygon.

III

$\triangle$	$\square$
4	3
5	$3\frac{1}{2}$
6	
7	
8	

Fig. 6

Teacher's Guide

As teachers prepare guidesheets for use in their classes, they should also prepare a teacher's guide for each guidesheet. The teacher's guide should explain the objectives of the activity and describe the materials to be used (telling how they can be constructed or purchased, if they are not already available). It can discuss alternate procedures, suggest key questions, and point out difficulties that the teacher might not have anticipated. It can also suggest ways in which the investigation could be extended, and suggest references for further investigation. Answers can be provided for some of the questions.

Teacher's Guide for Linear Measure (LM–1, page 69)

This guidesheet relates primarily to objectives 2, 5, 6, 8, and 11 listed earlier in this chapter under "Objectives in the Study of Linear Measure." The teacher should make sure the students understand that the drawings of the rulers in the guidesheet are not full scale. The adding machine paper should be a heavy grade. Students will probably need help finding a set of eleven parallel lines for dividing the unit. Since an ice-cream stick has round ends, students shouldn't try to divide it into equal fractional parts. Instead, they should mark an ice-cream stick length along the edge of a piece of paper and divide the marked unit. Some students will place the ruled notebook paper on top of the adding machine paper and will have trouble. The green and yellow sticks can be ¼-inch dowel rods 35 and 14 inches long, respectively. There will be some noise as students perform activities such as measuring the door.

Teacher's Guide for Functions: Parts of a Circle (F–1, page 71)

This investigation helps the student learn to make a prediction by using the graph of a function.

The chart must be prepared with care or the results will not be convincing. The borders of the graph paper should be removed so the markings can be aligned and uniformly spaced. You may want the students to use masking tape that can be removed without tearing the paper so the chart can be used more than once. Students must place each circular object on the chart so that its diameter will be marked from point 0 along the horizontal axis. The object should be traced with a sharp pencil so that the circle is as close as possible to the outline of the object. The pieces of yarn must be parallel to the vertical scale. The dimensions of the chart should be such that the piece of yarn representing the circumference (approximately 126 cm.)

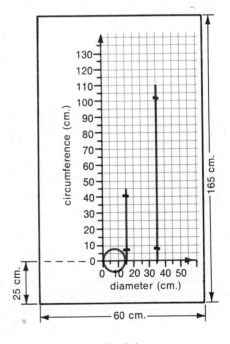

Fig. 3-1

of the circle with the largest possible diameter (40 cm.) can be accommodated. Therefore the chart must be about 15″ × 50″. Two poster boards end to end are 24″ × 76″. Enough margin must be left at the bottom to accommodate the circles that are drawn on the horizontal axis.

Students can be helped to make a generalization about the relations between the circumference and the diameter if they attach a piece of yarn to the chart at 0 and extend it along the tops of the other pieces. An answer like "The circumference is three times the diameter" is a reasonable guess for upper elementary school students. Depending on the group, the teacher may wish to amplify on this and give either 3.14 or $\frac{22}{7}$ as a better approximation to π.

Teacher's Guide for Functions: A Measuring Wheel (F–1a, page 73)

Although this activity requires sawing, drilling, and sanding and much preparation and cleanup, it is very worthwhile. It combines psychomotor development, mathematics applications, and development of a useful measurement tool.

The interpolation for finding the diameter must be fairly accurate, since the diameter measures that will give a circumference measure of 1 meter, accurate to the nearest centimeter, are between 31.7 cm. and 32.0 cm.

The wheel can be used to measure along curved paths. The accuracy of the wheel should be verified along a straight path. If it measures less than 1 meter in circumference, strips of masking tape can be wound around the rim to give the wheel added diameter. If it measures more than 1 meter, careful sanding and rechecking are called for.

Marks should be placed along the circumference of the wheel to allow for greater precision of measurement.

Circles with many different diameters can be drawn on a paved area and measured with the measuring wheel. The results should be either predicted or verified from the derived value of π.

Teacher's Guide for Pentominoes (P–1, page 75)

This is a problem solving experience in spatial perception. Success is *not* dependent upon a student's competence in computational skills or concept mastery. Students who excel in this may have traits of originality and creativity.

The square-inch pieces can be cut from poster board or purchased from an educational supply house. Students can be given sheets of squared paper (1 inch squares) on which to outline the various pentominoes. Then each pentomino can be cut out, so that no taping of edges is necessary. The twelve *different* pentominoes are shown on the following page.

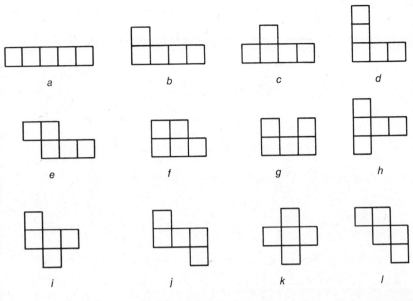

Fig. 3-2

Two pentominoes are different if one cannot be obtained by sliding, rotating, or flipping the other one. The mirror will help students discover which pentominoes could be formed from one another by flipping.

The wrapping exercise is very instructive. If the students are bright, they may be asked to tell why some pentominoes will not wrap a cube. Refer to figure 1: Those that won't wrap the cube are *a*, *d*, *f*, and *g*.

Teacher's Guide for
Shapes from Pentominoes (P–2, page 76)

All the coverings outlined on the guidesheet are possible. The answers are generally not unique. (The construction of the 3 × 20 is the only one we know to be unique.)

Hexominoes can be explored in a similar way. The foldings then are for a cube. *Polyominoes,* by S. W. Golomb, is a good source of ideas.

Another extension involves the use of cubes instead of squares. Each cube must be placed so it has a face-to-face contact with at least one other cube. Start with 3 cubes. Then use 4 cubes. There are two different arrangements of 3 cubes, 8 of 4 cubes, and 29 of 5 cubes. Most people who work with pentacubes, however, eliminate the 1 × 1 × 5, so that they work with only 28 (Golomb, p. 118). The arrangements can be used to form cubes (soma cubes). See Martin Gardner, *Mathematical Puzzles and Diversions,* book II, pages 65–77, for other ideas involving soma cubes.

When all the pentomino shapes have
been found, they can be cut out of
heavy paper for future use.

What shape rectangles can be made
by fitting the pentominoes
together?

Many interesting activities can be
done with pentominoes. Some
involve competition.

Teacher's Guide for
Attribute Cards (AC–1, page 77)

This guidesheet provides exercises in examining the properties of sets of objects. The formation of definitions and concepts is dependent upon one's ability to perceive properties of sets of mathematical entities. Size is a two-valued attribute of the cards (large or small); shape is a four-valued attribute (circle, triangle, square, or diamond); and color is four-valued (red, green, yellow, or blue).

These pieces can be cut from sheets of colored poster board or construction paper. Two pieces of like color can be glued together to give added strength. A more permanent set can be made from $\frac{1}{8}$-inch or $\frac{1}{4}$-inch plywood or pressed board that has been painted. A set of blocks (Attribute Games) can be purchased from the McGraw-Hill Book Company.

Charts I, II, and III should be prepared on poster board. More than one solution exists for exercises 3 and 4, which relate to figures 3 and 4.

Exercise 4 is a relatively difficult logical exercise for which the grade level can vary. Many variations of this game are possible.

One variation is a 36-piece, three-size, three-shape, four-color set of items. A set of pieces for one of the four colors is shown below.

Fig. 3-3

Make all the cards from pieces of cardboard of the same size. The three shapes are semicircle, triangle, and *squiggly*. Each should be done in four colors. There are three sizes of each shape. An extension of this activity involves having students place the pieces (as in exercise 3 or 4) in turn. Each is awarded one point for placing a piece. Anyone unable to place a piece misses his turn.

Teacher's Guide for The Geoboard (G–1, page 80)

The purpose of this exercise is to explore relations between the number of pegs touched. Students can determine the area of a figure by counting the squares and parts of squares enclosed by the rubber bands, thereby gaining experience with a lattice system. They also work with functions connecting numbers of boundary and interior points of a figure with the area of the figure.

A geoboard can be made by cutting a 9-inch square of $\frac{1}{2}$-inch plywood and driving 25 small nails into this square at even intervals (see fig. 3-4). The board can then be shellacked or brightly painted. Rubber bands should be sturdy and fairly large; string or colored yarn can be used instead, but neither is as easy to work with.

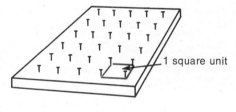
1 square unit

Fig. 3-4

In exercise 6, answer *c* is correct. In exercise 8, *a* is correct. After filling in table III in exercise 9, students should develop the number sentence $\frac{1}{2}\triangle + 1 = \square$. Students who are able to generalize from the results of exercises 6, 8, and 9 should arrive at the number sentence $\frac{1}{2}\triangle + \bigcirc - 1 = \square$.

If students show a great deal of interest in this activity on the rectangular geoboard, a similar activity on what we call an isometric geoboard (see fig. 3-5) can be explored. This activity would show how arbitrary the choice of unit is. The unit would be called a trian and the relevant formula would be $A = B - 2 + 2K$ (where A is the area, B is the number of pegs on the boundary, and K is the number of pegs inside the figure).

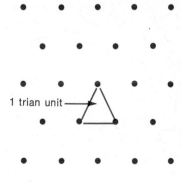
1 trian unit

Fig. 3-5

Helping Students to
Make Generalizations from Investigations

Some investigations develop mathematical concepts (see guide-sheet F–1). Others reinforce or extend concepts. In these experiences the student goes beyond manipulating objects and making observations. He must be led to make predictions, verbalize discoveries, make comparisons, and suggest modifications and extensions (see guidesheets F–1a, G–1, and P–1).

The guidesheet may elicit many of these responses by posing problems or questions (see guidesheets P–2 and AC–1); but the guidesheet has its limitations, for it must be short and appealing to the student. The guidesheet may sometimes help a student to evaluate his decisions through a built-in verification (see guidesheet LM–1) or a second method of attack. However, the guidesheet cannot adjust to each student's responses.

The teacher must visit each investigating team and engage the members in a dialogue aimed at probing the depth of their thought. The teacher should also plan to involve the entire class in discussions of this type. It is good to have a team report on its investigation, especially if that particular investigation was not made by all the teams. This report could be in the form of a display, a short film clip, a series of photographs, an oral report to the class, or an assembly program.

4

Facilities for the Laboratory Approach

Before the laboratory approach is implemented, it is necessary to take a close look at the design of the mathematics classroom. Whether a classroom is to be adapted or a new room set up, the space and facilities will strongly influence the success of the program. The laboratory should be planned to accommodate a variety of activities. Students will work as a class, in small groups, and as individuals. Sometimes they will sit as a discussion group; sometimes they will view the chalkboard, projection screen, or demonstration table; sometimes they will be working exercises in a textbook, reading, measuring distances, weighing objects, playing games, performing experiments, asking questions, and discussing what they have learned.

Space and Facilities for the Laboratory

The following standards for space and facilities can help make the laboratory approach more successful:

1. *More space per pupil than conventional classrooms.* In the mathematics laboratory much movement is generated in the taking out and returning of materials. Students move about and talk as they observe and manipulate these materials. Laboratory activities require more space than the activities of a conventional class. Use of the hallway and the school grounds for some activities can be considered.

2. *Flexibility in arrangement and use of classroom equipment.* To provide freedom in the organization of laboratory activities, it may be necessary to discard the practice of assigning a particular desk or table

A large room with carpeting and drapes is a good laboratory setting.

Students especially enjoy activities
that make use of the school
grounds.

The laboratory is more flexible
when students are not assigned
individual desks.

to each student. Level, flat-topped working surfaces will accommodate a wider range of classroom activities than will slant-top desks.

It is best to use folding tables and chairs that can be moved about easily and stacked in a small space when not in use. If the tables are square, rectangular, or trapezoidal, they can be placed together to form a larger working space for a team of investigators. Narrow tables (1'6" to 2' in width) are suitable for classroom use when all students are facing toward a focal point such as a screen, television set, or blackboard.

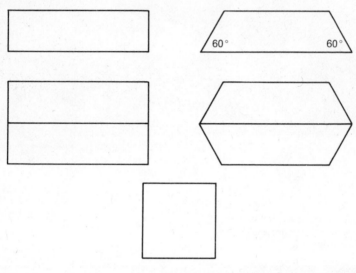

Fig. 4-1

A stationary table with a sink, a vise, and mounts for pulleys is very useful. It can be used for experiments by the teacher or a class demonstration team. Tables at which students stand to work (about 36 inches high) will provide welcome relief for students who become restless from sitting all day.

3. *Soundproofing of classrooms.* Carpeting on the floor, acoustical ceiling tiles, and draperies along walls are effective means of keeping the noise from student discussions and equipment like calculating machines at an acceptable level. Some schools use soundproofing rugs and draperies only in the work areas in which students are reading, viewing films, or using calculators. Soundproofing of carrels, division of working areas by cloth screens, and installation of bulletin boards will also help. An air conditioner provides a constant hum that softens distracting sounds. Background music can make noises arising from student activities less distracting, while improving the concentration and discipline of students.

Study carrels afford students some seclusion from the activity of the laboratory.

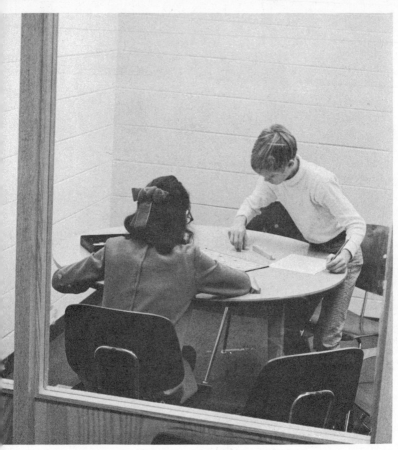

Too noisy? Private study rooms are
a solution.

4. *Work bays and study carrels.* Some privacy can be obtained
by use of room dividers such as cabinets and movable folding screens.
Folding screens can be made of pieces of pegboard, pressed board,
two-ply corrugated cardboard, or plywood tied or hinged together. The
screens can be left natural, painted, treated for a chalkboard surface,
or covered with flannel. They will deaden sound as well as provide
additional display area. A simple study carrel can be made from acoustical panels mounted on plywood. The carrel can then be placed on a
table to form a working bay. Figure 4-3 shows partitions 30″ × 30″ with
a backing 30″ × 84″ all mounted on a table 24″ × 84″. A shelf 12
inches wide is installed 15 inches above the top of the table. This
arrangement will accommodate three students. A simpler set of carrels
can be constructed for four students on a square or round table. The
panels should be 30 inches high and be as long as required to fit the
table. The carrels can be draped with dyed burlap to provide color and
a display area and to deaden sound.

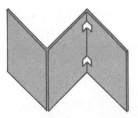

A folding screen to be used as a room divider

Fig. 4-2

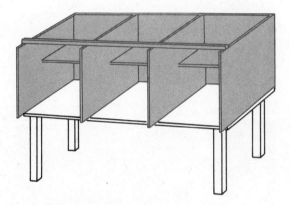

A set of three study carrels with a shelf

Fig. 4-3

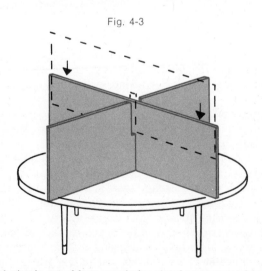

A simple set of four carrels for round or square tables

Fig. 4-4

Centers can be provided for quiet reading, listening to tapes, viewing films, computing, drawing, working with tools, and playing mathematical games.

Minor classroom renovations can be carried out with a minimum of expense if school staff members, students, and their parents are called upon to help.

5. *Technical equipment.*   A variety of technical equipment is being used in mathematics laboratories. Most of these laboratories have an overhead projector on a movable table. A dry-copying machine is useful for copying students' work, guidesheets, or order blanks from local businesses. Electric calculators can be used very effectively. (Manual calculators are satisfactory if electric ones are beyond the school budget.) A tape recorder is helpful in giving instruction to students who have reading difficulties. Some other useful items are a cash register, a gas pump indicator, filmstrip projectors, Reading Accelerators® (for timed drill), an 8mm cartridge projector, and a 16mm sound projector. A more comprehensive list of technical equipment is found in Appendix A.

How fast do you walk? A stopwatch has many interesting uses.

A machine that is simple to operate can be used by students to copy guidesheets and student work.

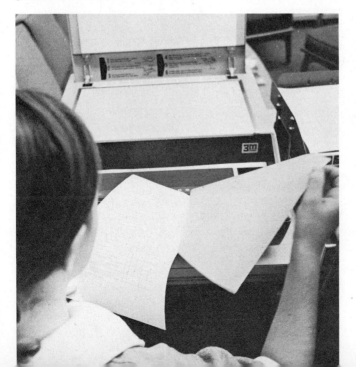

Working with a cash register helps
students see practical applications
of mathematics.

6. *Adequate lighting and electrical outlets.* There must be good lighting in the work areas, since many of the classroom activities require visual attention. There must be adequate electrical outlets—to handle the technical equipment that is now available as well as that of the foreseeable future.

7. *Storage for laboratory materials and student books.* During a laboratory period some students will be in work bays or study carrels, a team may be at a teacher demonstration desk, another at a counter along one wall, a third outside the classroom, and still others working at student tables placed together in pairs. Materials such as reference books, mathematical games, calculators, and tape recorders will be in their designated areas. Students will be using a variety of materials that must be carefully classified and stored in the classroom or brought into the classroom from a materials center on a cart.

Materials can be made available for student use in two ways. For some experiments it will be most convenient to set up stations at various locations around the room. Teams will then move from station to station performing an experiment at each. For other experiments materials can be placed in trays on a counter near the storage cabinets. Each team can then carry a tray of materials to its working location. Measuring sticks can be hung at two or three places on a storage cabinet or near the chalkboard, since these are used in most experiments and do not fit into the trays.

The materials must be carefully classified and placed on open shelves or in filing cabinets, closed cabinets, or map file drawers. Storage areas in all locations should be labeled to facilitate student use. Materials on the open shelves can be kept in boxes labeled in accordance with their uses. Open shelves for students' books should be provided near the entrance to the room.

Materials for the Mathematics Laboratory

The materials that a teacher needs when engaged in laboratory activities are many and varied. Materials are most valuable if they are simple, durable, accurate, and appealing to youngsters. Of course the materials must be suitable for use with interesting, challenging, and workable problems.

To be useful, materials must be classified, stored in clearly labeled containers, kept in good repair, and made readily available to the teacher and students. Some schools prefer a materials center with someone in charge of distribution. Carts for transporting materials to the various classrooms are supplied. Other schools keep a few of the more expensive and less frequently used items in a center, storing most materials in the teachers' classrooms.

The materials listed in Appendix A are suitable for use in mathematics laboratories for grades 5 through 9.

Plastic trays on rollers form a storage system for small objects and unfinished projects.

Learning mathematics is fun when
you have brightly colored sticks to
help you see number relationships.

Arrangements of Laboratory Facilities

Laboratory facilities can be provided for in several ways. Which plan a school should adopt will depend upon its answers to questions such as these:

1. *Is only limited or occasional use of the laboratory approach being planned?* Many teachers are looking for a fresh approach in their mathematics classes but do not want to commit themselves to extensive use of laboratory methods. They may see little evidence of their colleagues making effective use of these methods. They may want to break themselves in gradually to make sure they are able to maintain discipline with such a free and individualized approach. Before they have had much experience with the laboratory approach, they may think of laboratory activities as appealing to students and a good change of pace but may not be sure that serious learning takes place. To these teachers, the learning of mathematics is a process of organizing ideas; they feel that the best way to get ideas organized in the student's mind is to program and closely structure a discussion of the ideas. However, after trying the laboratory approach, most of these teachers are able to see its inherent structure; they see that interests and needs developed by students are structured according to what Tolman refers to as a cognitive map (Hill, pp. 120–121); and they become involved to a much greater extent. They become aware of the role diagnostic instruments and remedial devices play in the approach. They see that students involved in the laboratory become aware (however vaguely) of their own deficiencies. Finally they find that they have to structure their discussions with students more carefully (although perhaps more spontaneously).

The teacher who plans to make only limited use of laboratory methods should keep several things in mind. First, he may want a reorganization of the staff that will allow for a laboratory resource person. This person would be in charge of a room containing a set of laboratory materials, all of which he has used or with which he is familiar. He might take materials into the classroom, or he might set up a schedule for students to come to the laboratory. One disadvantage of having students come into the mathematics laboratory is that they may get the impression that the laboratory is not an integral part of the mathematics program.

Second, the teacher may want to use materials from the laboratory in a teacher demonstration. Teacher demonstrations cut down on student movement, but they also help to kill student initiative in that they offer few if any advantages over other techniques. However, this method can be useful in relating mathematical abstraction to reality when time or materials are short. A cart can be used to transport materials from the laboratory to the classroom.

Finally, the teacher may want to introduce activities seemingly unrelated to the unit of study. He may want to do this on a Friday or the day before a holiday. Examples are games, puzzles, and open-ended experiments.

2. *Are laboratory projects an integral part of the mathematics program, or are these projects optional and not necessarily related to the program?* If the first alternative is selected, the classroom should be equipped so that a variety of activities can be carried on simultaneously. Each classroom teacher would be in charge of the laboratory activities of his students. If the second alternative is selected, students can be sent to a centrally located mathematics laboratory to have the teacher in charge supervise and guide them.

3. *How large are the classes in which laboratory methods are being used?* If there are more than twenty-five students in a class, it may be advisable to have a large classroom and a helper for the teacher. This helper may be a mathematics student in an upper class, a housewife from the community, or an education student from a nearby college.

4. *Is team teaching in operation in the school?* Figure 4-5 shows a large classroom for forty-two students taught by two teachers. One teacher is responsible for organizing the laboratory materials, writing guidesheets, and supervising laboratory work. The other is responsible for collecting materials, assessing progress, and assigning remedial work. There are fourteen tables (1'6" × 6'6"), each suitable for three students, in front of the screen (for overhead projector) and the chalkboard. These tables are used in this position when the entire class is participating in the same activity, such as viewing a film, listening to a presentation by a teacher, or taking tests. When all the tables are not needed in this part of the room, some can be placed behind the movable screens and used for laboratory experiments.

In the laboratory side of the room there is space for a small conference room (for teacher counseling or student conferences), study carrels for six students, a table for four calculators, a typewriter with a mathematics keyboard, a large map, a file for posters (with a large paper cutter on top), and 36-inch-high cabinets along a section of two walls for materials to be set out on and for stand-up experiments.

Some students may be at work in the laboratory while others are involved in a class discussion in the other part of the classroom. In order for both types of activities to go on at the same, it is important that some soundproofing be used.

The floor plans and photographs at the end of this chapter are examples of laboratories that are in use today.

5. *Is a new facility being planned, or must the existing classroom be adapted?* Imaginative and careful planning of a new facility will greatly enhance its usefulness. Mathematics teachers, architects, and specialists in mathematics education should be brought into the planning.

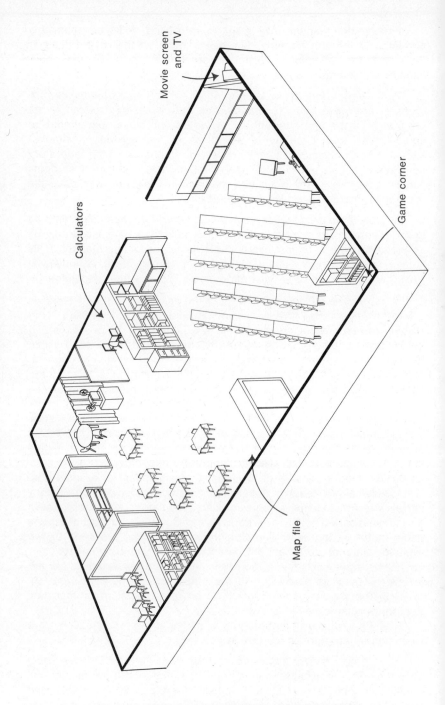

Fig. 4-5   Mathematics laboratory for team teaching
33 ft. × 54 ft. (42 students)

It is more likely that an existing classroom would have to be adapted. Figure 4-6 shows how a conventional 26′ × 30′ classroom could be converted into one that can be used for total class discussion or for work in small groups. Folding screens and soundproofed walls provide some privacy and quiet in three work areas: the computation center, the quiet reading center, and the mathematical games area. There are tables and twenty-four chairs for students. A teacher demonstration desk is located at the front of the room. A screen fastened to the ceiling can be lowered for use with the overhead projector. Space for a television set can be made available. Plans should include the soundproofing of walls and ceiling.

S_1, S_2, S_3, and S_4 are storage cabinets six to seven feet high. These have open shelves for storing materials in boxes or trays. Along one wall is a storage cabinet 36 inches high with a flat top for stand-up experimentation. Above this cabinet there could be a bulletin board about ten feet long, which can accommodate a large assignment record for the class (see chapter 2). On one side of the doorway should be shelves for students' personal materials; on the other side could be a display table with display shelves above.

6. *How many teachers will be using the laboratory approach?* If several teachers plan to use it, it will be advisable to have the department discuss the use of the materials. Expensive materials that are used only occasionally (transit, plane table, sextant, stopwatch, balance, thermometer, films, certain tools) should be in a centrally located mathematics center.

Figure 4-7 shows one of two adjoining classrooms and a shared area that can be used as an office as well as a room to house drawing board, typewriter, spirit-master machine, workbench, movie projector, large file for maps and posters, and three calculators. In the office space there could be filing cabinets and shelves for books, as well as a lockable cabinet for storing materials.

7. *How much money is available for the purchase of laboratory materials?* If the supply of materials is limited and if several teachers plan to use these materials, it becomes necessary to concentrate many of the materials in one room. One person would take charge of this room and devise a plan for the teachers to schedule the use of the materials. Many of the materials can be used in this room by students and by the teachers while making their lesson plans. The room should be centrally located with respect to the classrooms, and means for transporting the materials to the classrooms should be provided.

The following could be kept in a central mathematics room: a typewriter with a mathematics keyboard; a spirit-master machine; office supplies; a dry copier; a large map file cabinet for posters; a large (24″ × 24″) paper cutter; a drawing board; a film projector and films; calculators; reference books and booklets; transit, plane table, map projection globe, sextant; a workbench with tools (such as a hammer, a vise, a screwdriver, nails, and an electric drill); mathematical models;

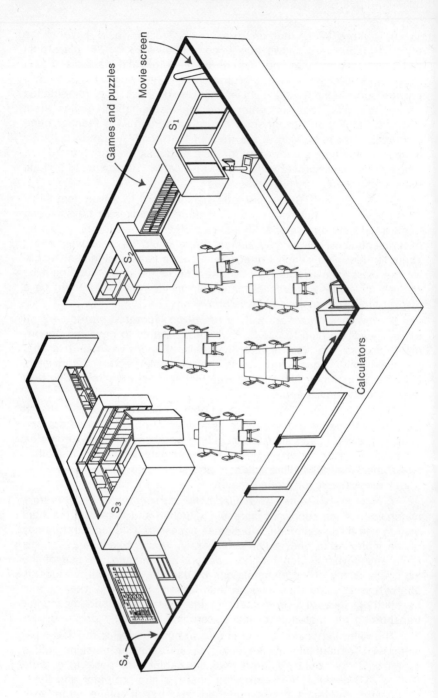

Fig. 4-6   Conventional classroom adapted for laboratory activities
26 ft. × 30 ft. (24 students)

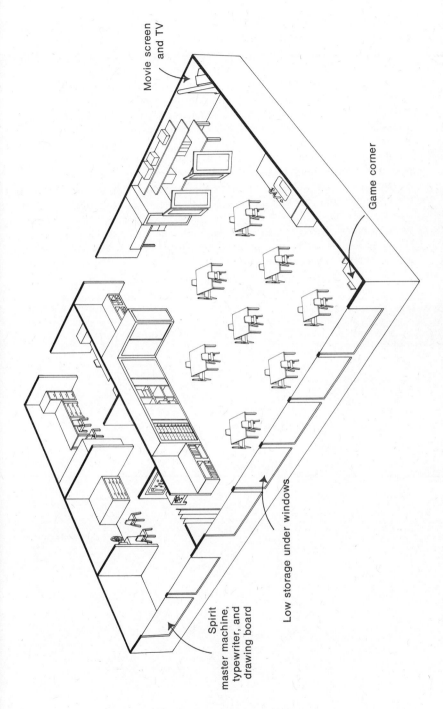

Fig. 4-7 One of two adjoining classrooms combined for laboratory activities

transparencies; devices for measuring (such as vernier calipers, micrometer calipers, balances, and thermometers).

Any planning for laboratory facilities must take into account the goals of instruction and the limitations under which the teachers operate. A careful consideration of the classroom space and facilities will do much to facilitate the achievement of the goals of instruction. The most effective arrangement of facilities depends on the particular circumstances. A spirit of inventiveness, improvisation, and experimentation must prevail. As a final suggestion, it seems advisable for a department wishing to initiate the use of laboratory methods to visit schools that are using these methods and to compare the various plans of operation. A teacher who initiates such a study should first consult his administration for guidance, and then prepare a questionnaire to be sent to schools in the vicinity. The questionnaire should be designed to determine to what extent those schools use the laboratory approach. He should then arrange for visits to these schools and prepare an action report for his department.

On the following eight pages are floor plans and photographs for four laboratories that were in use as the book was being written. The photographs are keyed to the floor plans.

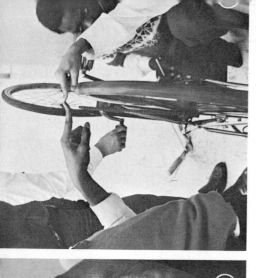

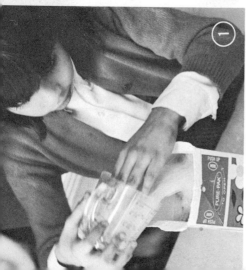

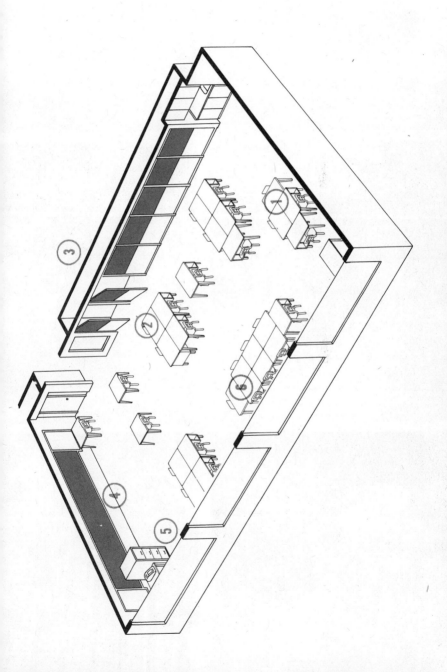

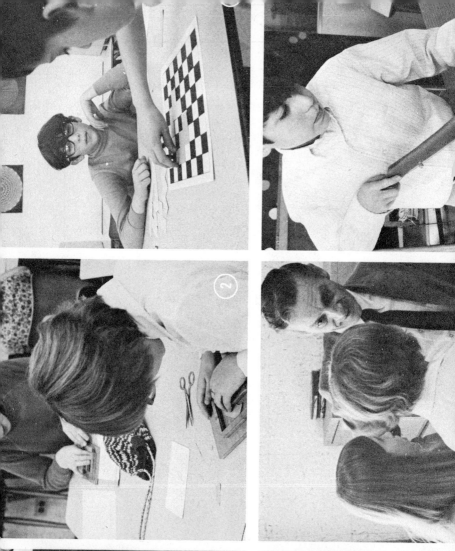
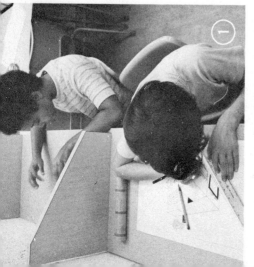

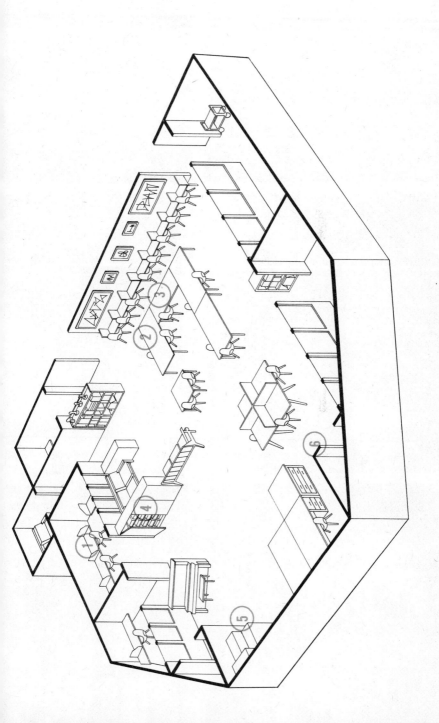

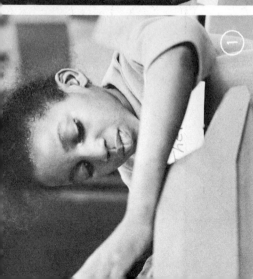

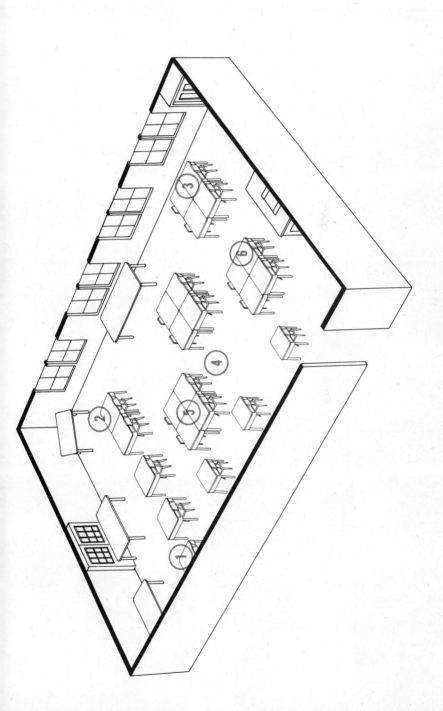

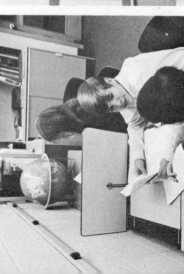

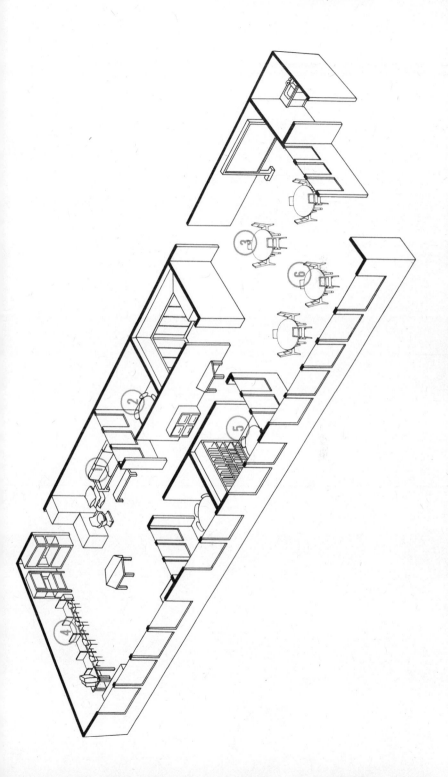

5

Assessing
the Effectiveness of the Laboratory Approach

It is difficult to move toward a goal unless you know your starting point and have some means of determining your progress. In education this principle can be expressed as follows: "Teaching begins and ends in evaluation."

The laboratory approach can make more use of the existing tools and methods of evaluation than can conventional approaches. The personnel of the school must learn to use these tools. They must learn how to completely determine what progress a student has made. They must decide whether it is meaningful to assess his progress on the basis of teacher-constructed tests alone, or whether a large percentage of his learning has been of the type that would be difficult to assess by means of such tests. They must decide how to evaluate the materials, the activities, and the teacher.

Theoretical Basis for the Laboratory Approach

Before the authors provide their recommendations, the ideas behind the laboratory approach will be reviewed and extended.

Psychologists, educators, and learning theorists have devoted years to the study of the processes by which learning takes place. Their findings have caused many to question traditional teaching methods and to search for something more in accord with the dynamics of learning. These ideas have led toward the laboratory approach and related teaching techniques such as those practiced by Maria Montessori in Italy and, more recently, by the British through their Nuffield Foundation. Maria Montessori followed the tradition of Pestalozzi and Froebel in indicating that learning comes from the manipulation of objects. British educators supported by the Nuffield Foundation harbor a like belief, but they have fit this belief into a philosophy of a core curriculum similar to John Dewey's.

Jerome Bruner is one prominent learning theorist whose ideas support the laboratory approach. His ideas on this subject can be summarized as follows (Bruner [1964], pp. 310, 313, 314):

> Any domain of knowledge (or any problem within that domain of knowledge) can be represented in three ways: (a) by a set of actions appropriate for achieving a certain result *(enactive representation)*, (b) by a set of summary images or graphics that stand for a concept without defining it fully *(ikonic representation)*, and (c) by a set of symbolic or logical propositions drawn from a symbolic system that is governed by rules or laws for forming and transforming propositions *(symbolic representation)*. . . .
>
> . . . The sequence in which a learner encounters materials within a body of knowledge affects the difficulty he will have achieving mastery. . . .
>
> . . . when the learner has a well-developed symbolic system, it may be possible to by-pass the first two stages. But one does so with the risk that the learner may not possess the imagery to fall back on when his symbolic transformations fail to achieve a goal in problem solving.

Bruner's statement suggests the importance of a teaching approach in which the three elements of this sequence are all provided for: the student must be given the opportunity to manipulate objects so that he can form images that refer to their manipulations; he should then be allowed to relate these objects to images, and he should operate on symbols with the images as referents; finally, he should operate on the symbols by means of established procedures without having to refer to an image or object.

The writings of Zoltan P. Dienes are similar, though more oriented toward specific examples of the concept that he calls *multiple embodiment*. Dienes stresses the student's need to explore several physical manifestations of a concept and to synthesize these experiences in order to form the concept.

Jean Piaget is a biologist and naturalist by training, a psychologist by profession, and a logician by avocation. He has written many difficult books, which are more accessible when interpreted by others. For example, Peter Wolff has synthesized Piaget's view of intelligence in the following manner (Wolff, p. 100; italics added):

> Piaget has formulated the development of intelligence as a unitary process pertaining to all aspects of mental development, including sensorimotor intelligence, perception, cognition, and affectivity. *All of mental development starts with motor activity,* all motor activity is intentional and object-directed, and the end result of all motor activity is the elaboration of stable mental processes (schemata) which are the basis of intelligent thought.

According to learning theorists, manipulating objects is a vital part of forming a concept.

Piaget therefore has concluded that manipulation of objects or the body must precede *any* mental development.

Proponents of the laboratory approach claim that the method not only will do a better job of teaching but will enhance the development of the child in ways that present approaches do not. They say that students will learn to work together, to apply their knowledge, to take care of classroom materials, and to take responsibility for their own learning. They believe this is so because the laboratory approach builds on success rather than punish failure, and it uses current theories of learning rather than ignore them. The main reasons other approaches have not developed such learnings are failure to recognize how to provide for their development and failure to build evaluative instruments to assess that development.

An informal meeting of college examiners at the 1948 convention of the American Psychological Association led to the publication of two extremely influential handbooks: *Taxonomy of Educational Objectives,* Handbook I: *Cognitive Domain* (1956) and Handbook II: *Affective Domain* (1964). These divided learning into three "domains"—cognitive, affective, and psychomotor—for which the objectives are defined as follows (Krathwohl et al., pp. 6–7):

1. *Cognitive:* Objectives which emphasize remembering or reproducing something which has presumably been learned, as well as objectives which involve the solving of some intellective task for which the individual has to determine the essential problem and then reorder given material or combine it with ideas, methods, or procedures previously learned. . . .
2. *Affective:* Objectives which emphasize a given feeling tone, an emotion, or a degree of acceptance or rejection. . . .
3. *Psychomotor:* Objectives which emphasize some muscular or motor skill, some manipulation of material and objects, or some act which requires a neuromuscular co-ordination.

Since this classification system accounts for all the kinds of learning expected of a student in the laboratory approach, it can provide a valuable tool for the teacher in planning the method of assessment. Its principal value will be to help the teacher pinpoint the areas he wants to assess and choose the best means of finding out what progress has been made within these areas.

This leads to several major questions: What are the most valid means of assessment? How should the results of such an assessment be recorded? What are the effects of these records on the student and the teacher? It is the intent of this chapter to suggest answers to these and to other questions asked in this introductory section.

Scope and Purpose of Evaluation

What is to be assessed and to what extent? When most educators think of evaluation, they think almost exclusively about the student. When they think of evaluating the student, they think only in terms of

Once you've solved a problem, you should be able to apply the same method to solve other problems.

The laboratory approach encourages students to share ideas and work together to solve problems.

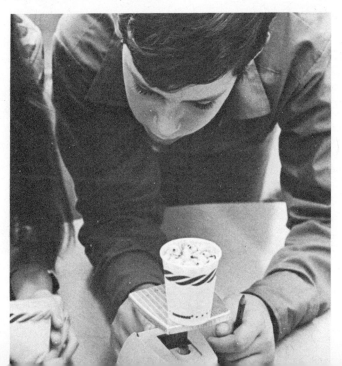

Students develop discipline when
working by themselves.

Curve stitching is a geometry
exercise that depends on the
student's psychomotor development.

teacher-constructed achievement tests. However, there are many other tools for assessing student progress, as well as many things to be evaluated.

The student must be evaluated at each level of representation: enactive, iconic, and symbolic. His development in each domain of knowledge (cognitive, affective, and psychomotor) must also be assessed. The tools that can be used vary. They include such things as observation forms, teacher-constructed achievement tests, and attitude inventories. The materials, the teacher, and the approach must also be evaluated. Therefore there must be tools for evaluating each of them.

The result of all evaluation should be improvement. We evaluate the student to find weaknesses and remedy them. We evaluate the materials, the teacher, and the approach for the same reason.

The intent of this chapter is to help teachers to better understand the scope and purpose of evaluation. The special relevance of evaluation within the context of the laboratory approach is explored in detail.

Instruments and Techniques for Evaluation in the Laboratory

How can the total learning of each student be assessed? In education, assessment is usually indirect. One person (usually the teacher) evaluates another (usually the student) through the use of some instrument. The various instruments should be continually evaluated and revised so that they are as effective as possible. Of the many evaluative instruments, the teacher-constructed test is the best known and most widely used. If such a test is used as a pretest, it can show whether a student is ready to begin work on a unit or whether he has already mastered the work to be covered. Some tests can be used to indicate how much a student can be expected to master in a given period of time. After a unit has been covered, a written test can be a good indicator of how much the student has mastered.

The teacher-constructed achievement test deals with objectives in the cognitive domain. An example is the posttest for the unit on ratio in Appendix B. At present this type of test is used mainly to grade students. It can be far more helpful, however, if it is used as a diagnostic instrument to show students what their difficulties are. For a test to be used in this way most efficiently, the items should be keyed to the unit of study. The following shows a keying of the test items from the posttest in Appendix B to the level-A guidesheets (item number to guidesheet number). If a student did poorly on test items 2 and 4, this would indicate that he had not mastered the work on guidesheet R-A-4.

Item	Guidesheet(s)	Item	Guidesheet(s)
1	R-A-1 & R-A-2	6	R-A-5 & R-A-6
2	R-A-4	7	R-A-4 & R-A-10
3	R-A-3	8	R-A-6 & R-A-10
4	R-A-4	9	R-A-7 & R-A-10
5	R-A-5		

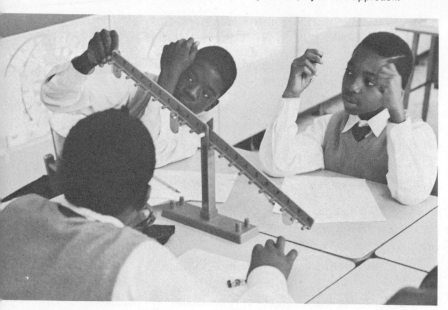

Another way of using such a test as a diagnostic instrument is the following:

1. Give a test that covers all the material, to discover where the student is weak.

2. Construct diagnostic tests that carefully define areas of difficulty. Areas should be further broken down to reveal what phase is causing the problem.

3. Write sets of problems that the student can do to help remedy these difficulties.

The following is such a diagnostic test for item 3 of the posttest in Appendix B, along with a set of problems on which the student can drill. The first test item has been set up to determine whether the student's problem is in visualizing the ratio or in filling in the picture when part of it has already been drawn.

Diagnostic Test 3 (Level A)

1. a. Draw a set B such that the ratio of the number of elements in set A to the number of elements in set B is (3:4).

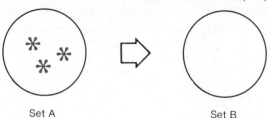

Set A   Set B

b. Draw a set C such that the ratio of the number of elements in set C to the number of elements in set D is (1:2).

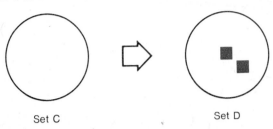

Set C   Set D

c. Draw sets E and F such that the ratio of the number of elements in set E to the number of elements in set F is (2:5).

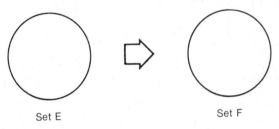

Set E   Set F

2. a. Draw a line segment *CD* such that the ratio of the length of segment *AB* to the length of segment *CD* is (2:3).

A  B

b. Draw a line segment *EF* such that the ratio of the length of segment *EF* to the length of segment *GH* is (4:3).

G  H

c. Draw line segments *JK* and *LM* such that the ratio of the length of segment *JK* to the length of segment *LM* is (1:5).

The following exercises would help remedy difficulties revealed by the diagnostic test.

Exercises 3-A

1. Draw two sets such that the ratio of the number of elements in the first set to the number of elements in the second set is the given ratio.

   a. (1:3)   b. (5:2)

2. Draw two line segments whose lengths are in the given ratio.

a.   (1:4)   b.   (5:3)

The pretest in Appendix B is another type of diagnostic test. It can be used to pinpoint deficiencies in cognitive development that would prohibit a student from making successful progress in a unit. After the difficulties are pinpointed, remedial help can be provided.

Used in this way, tests are invaluable in the classroom. They are, however, frequently used in a different way—namely, for grading. Not only does this not make full use of the test, but it is not entirely fair to the student. Since this kind of test rarely reveals the student's development in any but the cognitive domain, it does not give a complete picture of what he has accomplished. Also, although a teacher-constructed test can show where a student is failing to grasp the material, it says little about *why* he is having trouble. Is the problem really with the student? Or is it with the materials or the teacher? The suggestion is not that this kind of testing be eliminated as a means of evaluation, but that it must be supplemented by other means so that progress can be fully indicated. A change in emphasis—to its use as a diagnostic instrument—is also suggested. Many evaluative processes can help complete the picture of student progress. One of the most valuable is built into the laboratory approach. This is the guidesheet. The guidesheet and the resulting activity make it possible for the teacher to observe much of the learning process. He can therefore help the student before he goes too far off the track, and he can understand better how the student arrived at his conclusions. Observation also reveals what the student is learning in addition to the main objectives of the lesson. For example, a group of seventh-graders performed the activity described on guidesheet F-1 (chap. 3) under the direction of an inexperienced teacher. They set up their own chart by placing two pieces of masking tape at right angles on a large table and marking them off in centimeters. They then taped string to the table, having used it to measure the circumference of various round objects. When they had completed the activity, their teacher was convinced that all the students had learned that the circumference of a circle is a *little more than three times the diameter*. This was one of the main objectives of the lesson, and the students had discovered it for themselves with a minimum of guidance. They also worked well together—choosing objects from the classroom, measuring the diameters, guessing the circumferences, and checking their guesses. Furthermore, their conversation during this activity revealed that they had learned that they had to keep the same zero point for each value charted, that they had to keep the strings parallel in order to obtain the most accurate results, and that they had to measure as accurately as possible. The teacher

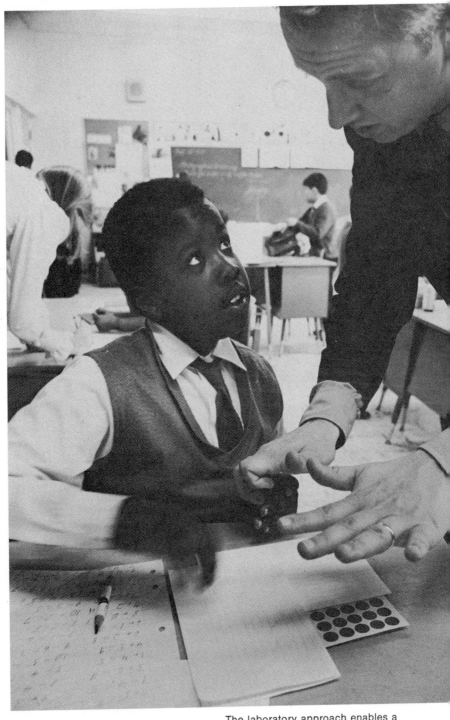

The laboratory approach enables a
teacher to see when a student needs
help.

Observing students at work, a teacher can guide them toward better use of their motor skills.

also observed that their psychomotor activity seemed normal and their attitudes were good. None of these had occurred to her as a thing to be assessed before the activity was started, but she observed them all. Most would not have been revealed by a written test.

Observation is possible in the laboratory approach primarily because there is so much to observe. As illustrated in the preceding example, attainment of the cognitive objectives can be observed. But that is just a beginning. Much of the student's development in the affective domain can be assessed—for example, his attitude; the degree to which he accepts or rejects activities, classmates, materials, and the teacher's help; his degree of attention; his interests and values; and his biases.

Since laboratory activities involve manipulation of materials and objects, they require the student to use his fine motor skills. Rarely do such things happen in the traditional classroom. Observation of the student while involved in laboratory activities can therefore reveal his psychomotor development, which probably is not assessed in school situations other than physical education, domestic arts, industrial arts, or handwriting classes.

How important is it to assess these other categories of objectives? Attitude, interest, and bias have a strong effect on a student's cognitive development. They cannot be dismissed by teachers with statements such as "All I care about is that they improve their scores on the achievement test," or "As long as they score well on the state test I'm happy," or "As long as they get good marks I won't complain." Attention must also be paid to development of the objectives of the affective domain.

Few schools have had their teachers develop objectives for assessment in the cognitive domain. Fewer still have had their teachers construct objectives for assessment in the affective domain, but much can be done to assess development in this area. Some ideas about assessment in the affective domain can be found in chapters 2 and 7 of *Evaluation in Mathematics* (Twenty-sixth Yearbook of the National Council of Teachers of Mathematics) and in the *Taxonomy of Educational Objectives,* Handbook II: *Affective Domain.*

Does he show interest in his work? His attitude will have a strong effect on his cognitive growth.

Several methods for assessing affective development other than teacher observation have been used by one of the authors while teaching in California. One of the simpler instruments is a mathematics ability check sheet (fig. 5-1).

Mathematics Ability Check Sheet

Student's name _____

                                         Last   First

Mathematics teacher's name _____

Check the appropriate box for each item:

	BELOW AVERAGE	AVERAGE	ABOVE AVERAGE
1. Ability to read and follow instructions			
2. Knowledge of the base-ten numeration system			
3. Ability to compute with whole numbers			
4. Ability to compute with rational numbers (in fraction form)			
5. Ability to compute with rational numbers (in decimal form)			
6. Ability to read, understand, and solve word problems			
7. Ability to reason			
8. Knowledge of plane geometric figures and their properties			
9. Knowledge of solid geometric figures and their properties			
10. Teacher's ability to understand your strengths and weaknesses			

Fig. 5-1

Each student completes the check sheet. At the same time the teacher completes his own sheet for each student. A subsequent analysis of the two forms and a conference can do much to help the teacher and student realize what they can do to help one another.

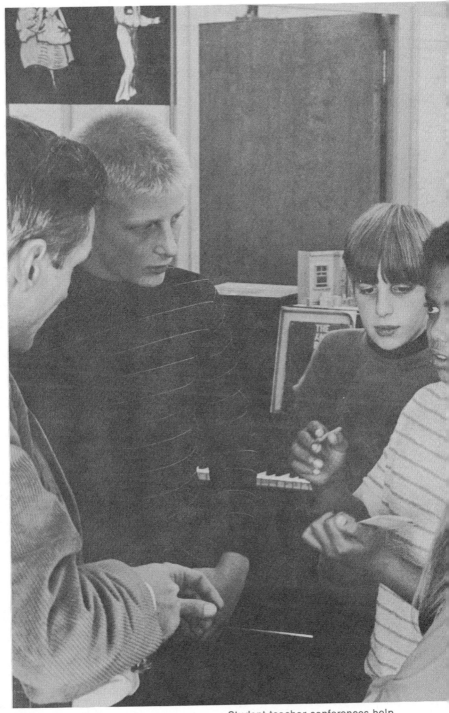

Student-teacher conferences help
assess development in both the
cognitive and affective domains.

Another tool for assessing student attitudes, biases, and concerns is the attitude inventory. Figure 5-2 shows part of an attitude inventory similar to one that is currently used in California. The construction and results of such an inventory will stimulate teachers to analyze themselves, the program, and the method.

Mathematics Attitude Inventory

Directions: Check the appropriate box for each item.

	ALWAYS	MOST OF THE TIME	SOME OF THE TIME	SELDOM	NEVER
1. Mathematics is my favorite subject at school.					
2. Homework has helped me to understand mathematics better.					
3. I worry about my grades in mathematics.					
4. I worry about the mathematics class I'm in.					
5. Mathematics teachers take enough time to explain the lesson so that I can understand it.					
6. I am afraid of mathematics.					
7. I think my marks in mathematics will be poor, so I don't try anymore.					
8. I think I would like to take more mathematics.					
9. I think I would like to be a mathematics teacher.					
10. Games help me do better in mathematics.					
11. Mathematics is a waste of time for me.					
12. My parents help me do my mathematics homework.					
13. My parents tell me that mathematics is important.					
14. I think homework does *not* help me do well in mathematics.					

	ALWAYS	MOST OF THE TIME	SOME OF THE TIME	SELDOM	NEVER
15. I think I would like school better if I didn't have to take mathematics.					
16. I have had good mathematics teachers.					
17. I use mathematics outside school.					
18. I like mathematics.					
19. My parents talk about how they use mathematics.					
20. I do my mathematics homework first.					
21. When I ask my parents for help in mathematics, they help me understand the problem.					
22. I do not try in mathematics.					
23. I think I will get a poor grade in mathematics.					
24. I like to compete with other students for grades.					
25. I read books on mathematics and find them interesting.					
26. I have been embarrassed in mathematics class.					
27. I am afraid to do problems orally or at the blackboard.					

Fig. 5-2

Combined use of a comment sheet (fig. 5-3) and student-teacher conferences is a valuable means of assessing both cognitive and affective growth. The comment sheet is a ruled sheet with spaces for names and several entry boxes beside each name (fig. 5-3). When combined with an elaboration of a key like the following, it can be a most effective device.

a. Poor study skills   *c.* Poor attitude
b. Acts pressured   *d.* Does not seem to hear well

e. Possible sight deficiency
f. Creative
g. Considerate
h. Not achieving as
   expected

i. Very effective in helping
   others
j. Constant source of good
   ideas
k. Fails to work well with
   others

Comment Sheet

School_____   Teacher_____

Class_____   Date_____

NAME	COMMENTS							

Fig. 5-3

A brief discussion with the student can often compensate when the teacher has not been able to observe him during a particular activity. For example, if the teacher in the circle-measuring activity described earlier had not been able to observe the students while they were working, she could have obtained much of the information by a brief discussion with the students, individually if possible, as soon as possible after they had gotten their results. The students should be encouraged to do most of the talking, prompted by questions that would reveal not only their understanding of the relation between circumference and diameter, but also their thoughts about the activity and the materials. Were they satisfied, for instance, that their measurements using cut yarn were accurate enough? Or would they have preferred some finer measure? Student-teacher conferences are most valuable when observation and questioning are combined with a report form and a guidesheet. The student is engaged in a guidesheet activity (with manipulative materials) that is at, or slightly above, his classroom achievement level in the given unit of study. While the student is so engaged, the teacher asks questions and fills in an observation form like that in figure 5-4. The teacher may also want to ask and make notes on questions such as the following:

1. Is there anything in this unit that bothers you?
2. Are you getting all the help you need?
3. Do you feel that you have learned as much as you should?

Any notes should be taken either during or immediately after the conference. *Do not depend on memory.*

It cannot be overemphasized that the student should feel comfortable during a conference of this type. To begin the discussion by

Student's name _____

Teacher's name _____

	EXCELLENT	GOOD	FAIR	NEEDS IMPROVEMENT	DATE/COMMENTS
Interest					
Self-directedness					
Concentration					
Attitude					
Use of equipment					
Care of equipment					
Health					
Appearance					
Speech					
Knowledge					
Motor skills					

Fig. 5-4

mentioning some class project that he did particularly well or some activity that he especially liked can help put the student at ease. Whatever means is used, it is important that he feel free to offer his opinions and to disagree with his teacher—that he be *himself*.

Items pertaining to objectives within the psychomotor domain appear in some of the preceding evaluative forms. Little has been done to investigate this domain, and little has been written except to stress that it is important. Learning theorists Bruner, Dienes, and Piaget follow the tradition of Montessori, Pestalozzi, and Froebel in saying that development in the psychomotor domain is inseparable from development in the other domains. It is vital to learning within the laboratory approach. The teacher should be aware of any limitations a student might have that could hinder his progress in laboratory activities. Parent-teacher discussions and observation are the best means of discovering any abnormalities in a student's psychomotor development.

The following chart (fig. 5-5) summarizes this section on assessment of the student. (Categories in parentheses receive less attention.)

Summary: Means of Assessing the Student

EVALUATIVE INSTRUMENT OR PROCEDURE	DOMAIN ASSESSED
Teacher-constructed achievement tests	cognitive
Guidesheet	cognitive, affective, psychomotor
Check sheet (ability)	affective (cognitive)
Attitude inventory	affective
Observation; comment sheet	affective, psychomotor (cognitive)
Student-teacher conference	cognitive, affective, psychomotor

Fig. 5-5

How can the materials be assessed?　This is probably the easiest question to answer. The materials to be evaluated are the guidesheets, the manipulative materials, and the tests. The first method of assessment is a checklist the teacher uses when reviewing the materials for use.

The checklist for the guidesheets should contain the following:

1. Are they readable?
2. Do they fit the objectives?
3. Can all the problems be done?
4. Is the language correct?
5. Is the mathematics correct?
6. Can the activity be completed in a reasonable amount of time?
7. Do the students have the interests and motor skills necessary to do the activity? Is the activity appropriate for their ages (mental, maturational, and chronological)?
8. Are the specified materials on hand, constructible, or obtainable?
9. Is the art appropriate?
10. Are the examples appropriate?
11. Is there humor when possible and appropriate?
12. Are all objectives attainable through use of the guidesheets?
13. Are the guidesheets attractive?

The checklist for the manipulative materials should contain the following:

1. Are the materials safe (nonflammable, edges smooth)?
2. Are the materials appealing (in color, size, weight, texture)?
3. Do the students possess attitudes and motor skills for successful handling of the materials?

Are materials fun to work with? Students should enjoy their laboratory activities.

4. Are the models enough like the real thing?
5. Are there other materials that could do the job better?
6. Are the materials durable?
7. Are the measurement materials accurate?

The checklist for the tests should contain the following:
1. Is possession of the necessary preskills assessed?
2. Is attainment of the objectives assessed?
3. Is the reading level appropriate?
4. Are the examples appropriate?
5. Do the items seem to be doing the job intended?

The second method of assessment is to observe and interview students involved with the materials. The same questions apply. Many teachers will be surprised by the ingenuity of their students in coming up with better activities and materials. Such interchanges improve everyone's attitude toward the entire teaching-learning situation.

Formal and informal discussions with other teachers to evaluate and discuss materials and their uses can be very helpful. Formal discussions can occur at workshops and professional meetings. Informal discussions can occur anywhere.

Finally, the teacher should be aware of materials of all types as he shops and attends conventions. Both experiences should be idea searches. Students should be encouraged to acquire this habit too. In this way the mathematics laboratory will continually be improved.

How can the teacher and the program be assessed? School administrators try to evaluate each teacher, but their attempts are usually infrequent and cursory. Therefore additional evaluation is necessary.

Who should perform the evaluation? This question leads to a pair of questions: Who knows the teacher and the program best? Who is most affected by the quality of the teacher and the program? Both questions have the same answer: the students and the teacher.

Many teachers are afraid to have students evaluate the program or themselves, and understandably so. It is difficult to open yourself and your work to criticism. However, it can be done with gratifying results if the program has produced changes for the better.

A good way to start might be by asking students to complete an inventory like that in figure 5-6, through which they can openly express their opinions of their teacher. After getting the general responses, the teacher should pick an area for improvement, such as fairness. To better determine what the students think about his classroom with regard to the chosen topic, he should then construct an inventory with the following characteristics. The items should be multiple-choice so that the students will not be inhibited from offering their honest opinion by a fear that the teacher will recognize their handwriting. The responses should be mutually exclusive. They should be specific as to the thing being considered and as to the teacher. Two possible sample items for assessment of teacher fairness follow:

1. My mathematics teacher
   a. helps each student when he needs help, except me
   b. helps some people more than others, at the expense of the others
   c. helps each student when he needs help
   d. helps me when I need help, but neglects others
2. When I have a problem, my mathematics teacher
   a. discusses the whole problem with me to see if my answer is right even though it is different from the one expected
   b. tells me what is wrong but has trouble explaining why
   c. tells me what the right answer should be but doesn't say what is wrong and why
   d. is often too busy to help me

Mathematics Inventory

Complete each statement below. (Two sample completions are given for each statement.) Give your *own* opinion. Be honest and serious.

1. My mathematics teacher has _____.

   Samples
   a. helped me understand mathematics
   b. not explained the subject so that I could understand it

2. In my mathematics class I _____.

   Samples
   a. get by with very little effort
   b. have worked very hard

3. I like mathematics, because _____.

   Samples
   a. it will help me later
   b. working with numbers is fun

4. I do not like mathematics because _____.

   Samples
   a. it is boring
   b. it is a hard subject

5. Mathematics homework is _____.

   Samples
   a. sometimes helpful
   b. never helpful

6. A mathematics teacher is the most inspiration to me if he

   _____.

   Samples
   a. is firm, fair, and patient
   b. doesn't embarrass students

Fig. 5-6

Is each student getting proper attention? Self-evaluation raises many questions for the teacher.

Student-teacher conferences can also help pinpoint teacher deficiencies and build rapport with each student.

Finally, the teacher should evaluate himself. The completion of a form like that in figure 5-7 will suffice if the teacher makes sure the items are pertinent to him.

There is an overlap in the various instruments and procedures for assessment. Often one procedure, such as a student-teacher conference, not only will tell about the student's development but also will reveal much about the materials and the teacher. Overlap can be helpful in spotting contradictions and in further pinpointing difficulties. However, care should be taken that overlap does not lead to over-evaluation, since this can end in boredom with the whole procedure, so that it becomes ineffective. It should not be difficult to avoid this and still cover everything that should be assessed.

Reporting

When reporting of student performance is mentioned, most people think of a report card with subjects along one axis, report periods along the other axis, and letter or numeric grades across the face. But what is reported on a report card? Are there other and perhaps better ways of reporting? Should reporting be synonymous with report card?

Teacher Self-Evaluation Form

ITEM	CLARIFICATION	COMMENTS
1. Objectives	a. Are the objectives of the unit clear? b. Is attainment of the objectives assessed by the test? c. Were the objectives appropriate to the group? d. Are the objectives meaningful to the students? e. Will attainment of the objectives be valuable to the student?	

2.
Student-teacher relationship
	a. Did I help everyone whenever possible?
	b. Did I favor some more than others?
	c. Did I praise good work?
	d. Did I give inappropriate criticism?
	e. Was each student clear about his responsibilities in the classroom?
	
	  1) Getting out materials
	  2) Putting materials away
	  3) Working independently
	
	f. Was I neat?
	g. Was I courteous and pleasant?
	h. Was I too gruff?	

3. Use of materials	a. Did I specify appropriate materials for the lessons? b. Did I give enough time to assessment of materials and methods? c. Were students who needed it assigned appropriate remedial work?	
4. Presentation	a. When addressing the class, was my voice clear? b. Were my statements and questions understandable? c. Did each student get a chance to participate? d. Were the students attentive? e. Were the students bored?	
5. Clarity of materials	a. Are all parts of the unit clearly related? b. Is the purpose of each part evident?	

Fig. 5-7

Teachers who merely write down a letter or a number to represent a student's development in school over a period of time probably have not thought carefully about what is being reported. What does his being above or below average for the class say, without a description of the class? Does the single grade take into account his development in the affective and psychomotor domains? Does it say anything about what might be hindering him from succeeding? If it represents percentage of work completed correctly, does a low grade indicate that the teacher did his best but is not good enough to help the student do better? If it represents effort, what grade should be given to the student who does perfect work with a minimum of effort?

It appears that single grades are meaningless. One alternative to this would be ordered quadruples in which the first component represents the average number of units covered by the class, the second component represents the number of units covered by the student, the third represents affective development, and the fourth refers to a key in which possible psychomotor deficiencies are listed. This procedure, however, is not entirely satisfactory.

Instead of adjusting the standard report-card grade to make it tell a more complete story, it would perhaps be better to eliminate the report card altogether. A more comprehensive verbal report to parents, followed by a statement of the teacher's opinion and a dialogue between the teacher and the parents, would be much more meaningful and useful. If time is not available for this, there are other things that can be substituted—for example, a written statement or a completed checklist representing what would have been said by the teacher at the parent-teacher conference. Either method should contain a statement of what action was indicated, what action the teacher was taking, and what action should be taken by the parents. In either case, all aspects of the student's development should be included. In conjunction with either method, reports such as that shown in figure 5-8 should be issued at the completion of each unit of study.

Throughout the year each student should keep a record of what he has done. This should include the following:

1. Guidesheets completed
2. Remedial or drill work completed
3. Chapters in textbook completed
4. Related outside work

The teacher should keep a folder for each student containing the following:

1. Teacher-constructed tests completed by the student that show his weak and strong points
2. Samples of the student's work
3. Notes taken during conferences with the student
4. Notes taken while observing the student
5. Reports of conferences with parents
6. Copies of reports issued to the parents
7. Notes describing any special talents or physical disabilities

Report of Student Progress

Student's name_____ Date_____

Teacher's name_____ Unit_____

	BELOW AVERAGE	AVERAGE	ABOVE AVERAGE
1. Use of class time			
2. Use of independent study time			
3. Mastery of unit objectives			
4. Class attitude and behavior			
5. Psychomotor skills			

Fig. 5-8

The material selected should give an accurate picture of the student, show how far he has gotten, and indicate any special help he will need. At the end of the year the record kept by the student should be added to this folder. This can then be passed along to the following year's teacher to help him see what type of help each student is going to need. Each year the new teacher should cull out unnecessary material; otherwise the folder will become too large.

In conclusion, let us ask the following questions: Is assessment a necessary part of the education process? If so, what is its role? There is little doubt that assessment is necessary and valuable. In order to teach someone, you must first know what it is that they need to learn; and while teaching them, you must have some means of determining how much they are actually learning. This is the diagnostic value of assessment. Is this its entire role? Many teachers have assigned it the additional role of *motivating force*. This is the use of the grade as a *reward* or *punishment*. Does a grade have value in this sense? Or is this a dangerous aspect of grading? There is a great deal of evidence to support the latter, and it is the students who already have learning problems who seem to be hurt most by the discouragement of a low grade. This is especially unfair, since it is debatable whether the student or the teacher actually deserves the low grade. When a student is given a top grade, the teacher, in effect, may be

To help keep track of their progress, students record activities as they are completed.

A notebook of work samples is
a big help when it is time to review
a student's work.

saying, "This student has done as well as, or better than, I had hoped." And when a low grade is given, he is really saying, "For some reason this student did not learn. I failed to help him overcome his problems; the approach I used to present this material did not help him to learn it."

The evaluation process, then, is vital to the teaching process, and traditional grading processes must be as carefully reviewed as traditional teaching processes. All parts of the teaching process must be designed so as to do the maximum to help each student learn. Any part of the process that does other than this must be carefully restructured.

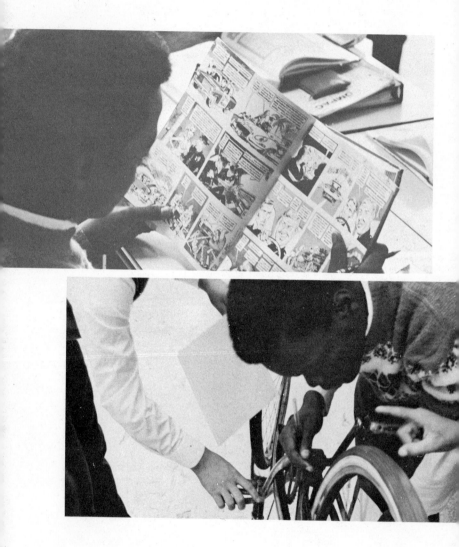

6

Mathematics for the Low Achiever

The 1963 norms for the SRA Achievement Series show that up to 17 percent of the junior high school population sampled were two years or more below grade level in mathematical competence. One estimate is that 25 percent of the school-age population are problem learners (Johnson, pp. 3–4). The terms used to define this population are varied. We will refer to the whole group as *low achievers*. We also use this term in a more limited sense, but the context should prevent confusion. Many low achievers will be school dropouts. The rest may continue in school but remain incompetent in mathematics. All face a work world that needs few unskilled workers but demands increasing numbers of semiskilled, skilled, and professional workers. Approximately one-eighth of the semiskilled and skilled workers in industry are in a retraining program to develop new skills; but we must have more retraining, and much better education in mathematics, if we are to eliminate the paradox of having jobs without people to fill them while there are people without jobs. According to H. L. Phillips, a specialist in mathematics with the U.S. Office of Education,

two-thirds of the skilled and semiskilled job opportunities on the labor market today are not available to those who lack an understanding of the basic principles (and skills) of arithmetic, elementary algebra, and geometry. Basic mathematical understandings are also essential to adult retraining programs for the unemployed. (Woodby, p. 2.)

It is vital to the well-being of our society that low-achieving students in mathematics become useful citizens. We must prepare them to enter the work force unthreatened by unemployment. The task of providing suitable mathematics programs and effective approaches to teaching low achievers is one of the most serious and persistent of our current academic problems. This problem must be faced by elementary and secondary schools as well as by community colleges.

Only sporadic attempts have been made to deal with the problems of low achievers in mathematics. We must look carefully at these students. We must discover how they behave and how they see themselves in relation to school as a whole and to mathematics in particular. We must ascertain the causes of their low achievement and analyze their needs. We must develop courses suited to their capabilities, backgrounds, interests, and needs. We must experiment to find new methods of teaching that are more effective with these students.

Characteristics of the Low Achiever

Figure 6-1 shows the different types of students who have difficulty learning mathematics. Certainly these students cannot fit comfortably into any one category. They are described by many terms, most of which are not flattering.

Many low-achieving students are *culturally disadvantaged* or *culturally deprived.* They are handicapped by an impoverished home background. Poverty, insecurity, and fatalistic attitudes of parents contribute to the development of low self-esteem, depression, and pessimism. Crowded living quarters and disorganized home life lead to the development of poor work and study habits. Because parents are not verbally oriented and there are few books at home, children have trouble acquiring skills in oral and written communication. Children of migrant workers have the additional problem of frequent change of residence, causing poor attendance and gaps in their education.

These culturally deprived children and their parents regard themselves as second-class citizens in the school community.

> From the classroom to the PTA they discover that the school does not like them, does not respond to them, does not appreciate (nor understand) their culture, and does not think they can learn. (Riessman, p. 55.)

The culturally deprived child is usually classified by his teachers as a low achiever or slow learner. It is partly to salve our consciences that we label him rather than accept him as he is and try to understand him, enhance his self-concept, and find approaches that will be successful with him.

The culturally deprived child is typically a *physical learner* (Riessman, p. 52). He has difficulty in grasping a concept unless he can visualize or handle objects that relate to the concept. The physical learner usually has a single-track approach to problems; he has difficulty in brainstorming ideas or making conjectures. He is slow in performing intellectual tasks involving symbolic thinking and generalizing.

Some low achievers underachieve only in mathematics. *Under-achievers in mathematics* are achieving in this subject significantly below what they are achieving in other academic areas. Many of these students have potential for much greater achievement in mathematics but for some reason have lost interest in this subject and have developed feelings of inadequacy. Research suggests that underachievers have feelings of high anxiety and generally come from homes where stress is placed on good grades (Small, p. 31).

Students Having Difficulty Learning Mathematics

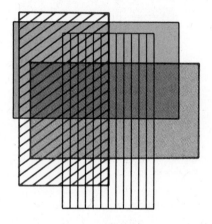

Low achievers in mathematics—those whose achievement in mathematics is two years or more below grade level

Underachievers in mathematics—those whose achievement is lower in mathematics than in other academic areas by one and a half years or more

Slow learners—those whose IQ is less than 90

Culturally disadvantaged—those whose environment is not producing the experiences necessary for normal growth in school

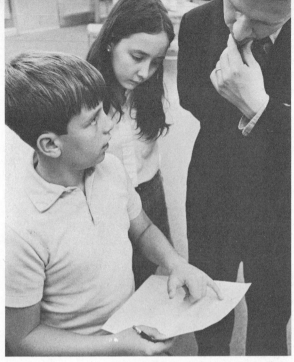

Self-confidence comes from having
your ideas accepted by your teacher
and classmates.

The underachiever displays little enthusiasm for mathematics. His attention span is short, and he frequently tunes out the mathematics teacher. In other classes, however, he often demonstrates the interest, perseverance, and participation that are missing in mathematics. Some underachievers have gaps in their mathematical background that have never been diagnosed and corrected. Another cause of an underachiever's lack of interest is that he seldom has enjoyed doing mathematics. Enjoyment, to him, comes from such things as (1) making discoveries; (2) having an idea *accepted* by the teacher or class; (3) doing things, other than drill exercises, that he is proud to talk about outside the classroom; (4) being complimented for contributions and achievements; and (5) setting up a goal and achieving that goal.

It is estimated that *slow learners* constitute 15 to 18 percent of the school population (Johnson, p. 9). With the exception of a few who achieve skill and self-confidence in computation, slow learners have difficulty in learning mathematics. Most of them have a reading level of three years or more below their grade norms; their communication skills are often very low. They are slow in discovering patterns, and they have trouble relating mathematics to various verbally described situations. It is often difficult for them to make comparisons, to give reasons for steps taken in a mathematical process, to organize ideas, or to brainstorm new ideas.

Slow learners prefer to work with objects rather than with abstract ideas, words, or symbols. For this reason, they have most success with such mathematical topics as statistical graphs, mathematical designs, measurement, and intuitive geometry.

Among the different types of low achievers, the slow learners are probably the happiest. They have learned to live with low achievement, they face less parental pressure than the underachievers, and mathematics does not appear as a threat to them. Many in this group are cooperative and dependable, and they often volunteer to do nonintellectual chores in the classroom.

There are also low achievers who do not seem to fit into any of the three categories described above. These are the *immature.* They lack goals and have emotional problems. These students usually have poor study habits in school. When the teacher makes assignments several days in advance, they are not careful to get the assignment and plan for its completion. Some completely ignore the assignment; others at the deadline date make excuses for not having it completed and ask for more time to work on it. Since they are disorganized and without specific goals, these students waste time in class. They have difficulty keeping track of their belongings. They seem reluctant to learn, are often rebellious of authority and impulsive, and appear to be pleasure-oriented. They seldom participate in large groups and are

Slow learners are often successful in making mathematical designs.

often uninterested in school activities. They rarely become group leaders, but they often become the class clowns and seek to gain attention through diversions.

Suggestions for Teaching Low Achievers

What can be done to develop a suitable curriculum for low-achieving mathematics students? What teaching methods and activities will help to capture their interest and motivate them? How can the mathematics teacher guide their learning? The solutions to these problems are not simple, but many of them can be found within the laboratory approach. The following suggestions incorporate the experiences of many effective teachers.

Study the student. Far too often, teachers, after identifying the uninterested students and low achievers in their mathematics classes, attempt only to control them. These teachers can expect to make little progress until they make an effort to understand these students. This means much more than looking at IQ and mathematics achievement scores. They must try to discover how each student perceives himself with respect to mathematics and to school in general. They must determine answers to the following questions: What is his home background? What are his work and study habits? his emotional traits? his reading abilities? Can he comprehend more than he can express verbally? What are things that he can do well? What are his interests? How does he spend his time?

It is also important to use diagnostic tests to determine each student's levels of comprehension and communication of mathematical concepts. These tests should reveal the student's level of facility with abstract mathematical symbolism. They should also reveal the extent to which various mathematical ideas are meaningful to him.

Enhance his self-image. Helping each student gain a more positive view of himself in relation to school and to the study of mathematics should be one of the principal goals of teachers of low-achieving students.

> Gradually it is becoming clear that the difficulties which people experience in most areas of life are closely connected with the ways in which they see themselves and the world in which they live. There is considerable and increasing evidence that student failures in the basic school subjects, as well as misdirected motivation and lack of commitment characteristic of the underachiever, the dropout, the culturally disadvantaged, and the failure, are in large measure the consequence of faulty perceptions of the Self and the world. At the elementary level it now appears that children's difficulties in basic academic skills seem to be a consequence of their beliefs that they cannot read, write, handle numbers, or think accurately, rather than of basic differences in capacity. Many students have difficulty in school not because of low intelligence or poor eyesight, but because they have learned to see themselves as incapable of handling academic work. (Purkey, p. 3.)

Does he have any special talents? Often they can be used to help him solve his learning problems.

Some ideas are not hard to understand but are very difficult to put into words.

There are probably very few people
who are not low achievers at
something.

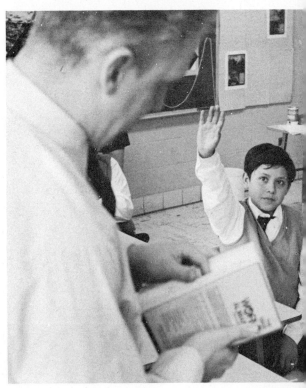

How big is a basketball? If he knows, he will be proud to tell the class.

A teacher must express genuine concern for each student and accept and respect him as he is. He should be careful not to let the student's learning difficulties lower his opinion of the student as a person. While most of us are high achievers in certain skills and concepts, there undoubtedly are some areas of endeavor—for example, speaking a foreign language, water skiing, or playing bridge—in which some of us would be classed as low achievers. Certainly we desire the esteem and approval of our associates even though we have difficulty with various endeavors.

Each student should frequently be given opportunities to make accepted contributions. This can involve supplying some item of information such as the official size of a basketball or the number of "cubes" in a certain sports car, preparing a piece of laboratory equipment, making a design or drawing, estimating a distance (often the low achiever can do this well), or solving a puzzle. Some teachers have adopted the practice of frequently using a student's *name* in giving him recognition for class contributions or achievements. This can be done verbally as well as in written form.

Displaying work that is well done is one way to give recognition to a student's efforts.

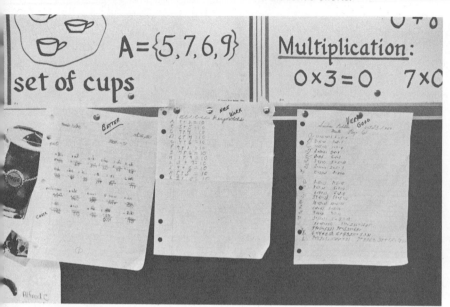

Each student should be helped to set up short-range, realistic goals and should be encouraged to believe that the goals are *attainable.* It may prove desirable to have each day's assignment on a looseleaf sheet or in a small booklet, rather than as part of a large, formidable-looking textbook. The student can be prevented from feeling handicapped by deficiencies in verbal communication if excessive wordiness can be eliminated from written materials, if technical words are kept to a minimum, and if complicated sentence structure is eliminated. Illustrations and pictures should be used, and reference should be made to everyday objects. A student who is handicapped by deficiencies in computation may be encouraged if he is allowed to use an electric calculator as a slave to do his computation.

Teachers must correct and give suggestions to students who are having difficulty in learning mathematics, but this must be done without discouraging them. Much more can be accomplished with these students by words of commendation and encouragement than by criticisms and red ink.

A person's self-image is enhanced when his class as a whole gains *status* and a sense of achievement. Interesting things should be happening in the classroom—things the students are talking about with pride outside the classroom. A feeling of belonging and of achievement may be promoted through a sort of Math Olympics in which there is competition between classes (possibly in different schools).

Teachers should give careful thought before adopting a no-homework policy for these students. Such a policy may be considered an admission that the class is substandard. It may be wise to plan assignments for each student that are useful, attainable, and challenging.

Develop a mathematics program that is relevant to the students' environment and interests. Mathematics content and classroom activities should be related to the students' experiences and to events that are of interest to them. Data collected from newspapers, students, businesses, organizations in the community, and school activities can be used to develop meaningful and significant projects from which mathematical concepts can be developed. As each topic of study is approached, the teacher should plan for students to obtain informal experiences with objects prior to the more formal symbolic treatment of the related mathematical ideas. Additional or more extensive compensatory experiences of this nature should be planned for the culturally disadvantaged students.

Provide a classroom climate to stimulate curiosity and participation. The classroom should be a place where students are trying to find answers to interesting questions. How can students be confronted with questions that are interesting to them? How can they be encouraged to seek answers? There are no simple answers to these questions.

The teacher should encourage students to ask questions. Often a teacher who understands his students can actually prompt students to ask questions they are interested in investigating. In a class atmosphere of mutual interest students will soon learn that questions that are significant to them will not be ridiculed or tossed aside by the teacher.

The teacher should ask questions. Questioning may be done to determine how well his students can carry out a process of mathematics, or how well they understand some concept. Questions of this type, however, will seldom make his students curious about mathematics. He must plan to ask questions that will cause students to wonder. Questions phrased in ways such as the following will help:

What would happen if . . .?

I wonder . . .

Can anyone show that Tom is on the right track?

Here are five digits—1, 2, 3, 5, 7. Each of you form a numeral, using these digits. Betty's numeral is 57,123. When we divide 57,123 by 9, the remainder is 0. What is the remainder when you divide your number by 9? Why do you suppose that *each* of you got a remainder of 0?

Silent or shy students usually become active in a game of asking questions. You may start by stepping out of the room and letting your class select some mathematical entity—for example, a whole number with two digits. You then try to identify this number in as few questions as possible. You must ask questions that can be answered by yes or no. For example:

Does she have a question? It is
important that she feel comfortable
asking it.

Is the tens digit larger than the ones digit? (No.)
Is the sum of the two digits less than 9? (Yes.)
Is the sum of the two digits greater than 5? (Yes.)
Is the number even? (No.)
Is the number greater than 20? (Yes.)
Is the number divisible by 5? (Yes.)
Is the number less than 30? (No.)
Is the number 35? (Yes.)

The class can then be divided into groups. The *asking* group leaves the room while the rest of the class selects a number, a property of a geometric figure, or a geometric figure (for instance, a 3-4-5 right triangle). The asking group then returns and proceeds to ask questions, trying to choose questions that will identify smaller and smaller subsets to which the entity belongs. At first many students will ask questions in a random, unplanned way, but as they gain experience they will develop more effective strategies of selecting questions.

Students should be encouraged to make conjectures or guesses. They may wish to classify them as *wild guesses* or *educated guesses.* It is often possible to have the entire class vote on two or three answers or guesses to questions. For example, if alternatives A and B have been contributed, students may be asked to close their eyes and vote by raised hand on *A, B, Neither,* or *Don't know.*

A	
B	
Neither	
Don't know	

Fig. 6-2

Low-achieving students often have difficulty in talking about their discoveries. Rather than insist that students verbalize, the teacher should ask them to *use* their ideas in a nonverbal way. For example, a student might see that the following exhibits a relation but not be able to describe what he has found:

$$1 = 1$$
$$1 + 3 = 4$$
$$1 + 3 + 5 = 9$$
$$1 + 3 + 5 + 7 = 16$$

He might be asked to use his rule to find the following:

$$1 + 3 + 5 + 7 + \ldots + 15 = ?$$

Plan a variety of activities. Plan two or three kinds of activities each day. Each activity should help the student achieve a specific objective, and adequate time should be provided for him to achieve the objective. He should be involved, not just an onlooker. The class may recount experiences together. Students may make conjectures, discover patterns, define problems, select methods of attack, read textbooks, and work exercises together. For the most part, however, students should work individually or in small groups. These individualized activities will include playing games, manipulating materials to find answers to questions, selecting a topic of special interest to investigate, reading textbooks and special-interest leaflets, using an electric calculator, taking diagnostic tests, doing drill exercises, solving problems, and preparing displays.

Varying the laboratory activities
increases students' interest.

A great deal of thought must be given to supervising students in a wide variety of activities. Materials must be readily available, specific rules of operation and behavior must be established, guidesheets must be on hand wherever needed, and the arrangement of classroom equipment must facilitate the successful pursuit of these activities.

Help each student to set up a program of self-improvement. Each student should have a record book or folder that remains in the classroom. In this he should keep the results of diagnostic tests, statements of goals, and records of achievements and remedial work. He should have the benefit of the teacher's frequent evaluation, guidance, and encouragement.

When needed, special instruction should be given on how to read a mathematics book, how to use an index, and how to locate specific information. Reading materials should be simply written and should include drawings or photographs to illustrate the more technical terms. Objects should also be used to aid student understanding of various concepts. Special attention should be given to the development of a mathematics vocabulary. The student should also be given instruction in how to plan his work or study period and how to check his results and locate his errors. In addition to this, he must be given instruction in how to express ideas correctly in oral and written form.

It is hoped that each student, gradually becoming independent of the teacher, will use more initiative in pursuing his activities and assume more responsibility for his own learning.

Establish standards of conduct and classroom routines. It has been suggested that students be encouraged to respond to questions, make decisions, raise questions, and work alone or in small groups. It is important that standards of conduct conducive to learning be clearly established and followed in an atmosphere of mutual respect. Routines must be established for participating in class discussion, scheduling activities, using classroom equipment and supplies, and working in small groups.

The Value of the Laboratory Approach for Low Achievers

The laboratory approach to teaching mathematics is of special interest to teachers who are planning programs for low achievers. It has received the strong support of both the Committee on Mathematics for the Non-College-Bound of the National Council of Teachers of Mathematics and the Washington Conference (1964) on the Low Achiever in Mathematics. Among the six guidelines for teaching mathematics to the low achiever reported by the conference were the following:

Particularly for the low achiever, the need for mathematics comes from experiences in the physical world.

A laboratory setting is especially effective for low achievers.

Evidence from research in psycho-pedagogies clearly indicates that active experimentation in which the child handles concrete objects and observes what happens precedes the formal operation stage in learning mathematical ideas. For slum children who come to school with a paucity of experience with manipulation of objects, the elementary teacher must provide the first selective planned environment in which active sensory experiences can take place. Only after the codification of experience can the real search for structure begin. (Woodby, p. 92.)

For what reasons do we consider the laboratory approach to be so valuable for the low achiever in mathematics?

It provides selective experiences with objects from which these students can develop concepts. As mentioned above, there is evidence from research that indicates that active experimentation in which the child handles concrete objects and observes what happens precedes symbolization of mathematical ideas.

Many low achievers have had inadequate experience in manipulating objects to find answers to questions. One can hope to accomplish little teaching of mathematical formalism until compensatory experiences have been provided. The laboratory is a planned environment with objects to be observed and manipulated.

The modern mathematics curriculum has been designed for middle-class students with average or above-average levels of verbal and symbol communication. Much of the content has no relevance to the daily life of the low-achieving student. The laboratory approach is an effective way to lend life and meaning to the abstract symbols and processes of mathematics. It is through relating mathematical ideas to his environment that the student will learn to appreciate the significance of mathematics.

It provides a means of communication that they can understand. Effective communication is necessary for effective teaching. Communication, however, may take place at various levels of abstraction. Some students are seriously handicapped in mathematics programs because they are able to deal with mathematical ideas only on the object level. These students have difficulty interpreting the usual type of textbook problem, in which a situation is described with words and symbols and questions are raised.

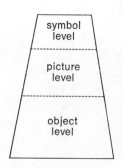

Fig. 6-3

The emphasis of the laboratory approach is on observing and handling objects rather than on manipulating symbols. Therefore it tends to open up communication lines to the low achievers. When objects and

Concrete experience must precede
abstraction of any concept.

A laboratory should contain a
variety of materials to observe
and manipulate.

the work with them become referents of words and symbols, youngsters
are helped to progress toward a more abstract level of communication.

**It provides an opportunity for them to become success-
fully in activities.** Another major handicap of the low achiever is his
lack of self-esteem. He has seldom experienced success in the traditional mathematics classroom. His embarrassment over continual failure
and his assumption that he will always fail inhibit his participation in
class activities. He is reluctant to explain a solution to the class if he
believes that most of his classmates are more familiar with the problem
than he is. On the other hand, consider the boost to his ego to be asked
to explain an investigation to the class when members of *his* investigating team are the only ones to have completed it!

When working in a laboratory setting, the low achiever may gain
some measure of confidence and sense of achievement for the first time.
With some practice, he may gain skill in estimating, solving puzzles, and
making predictions, since these activities often do not require symbolic
thinking. He is not under threat of failure when he punches the keys of
the calculator or times the bicycle rider. He is involved and he is having
some success; he does not seem hopelessly lost for long. He can make
mistakes and learn from these mistakes. When electric calculators were
introduced to students in a general mathematics class, one youngster
became overjoyed and began to jump up and down. A fellow student

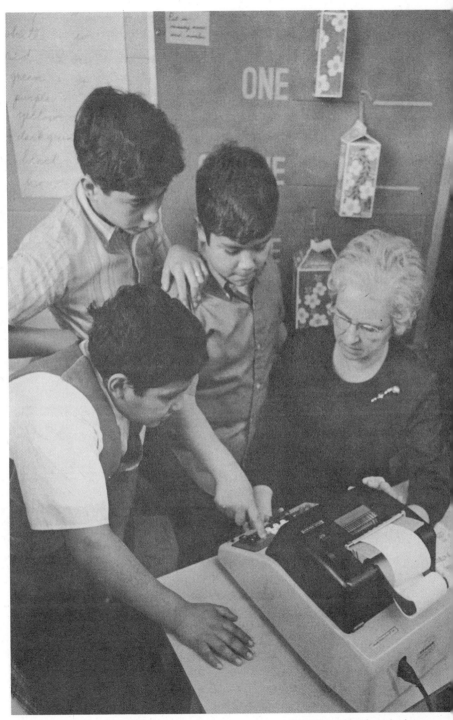

Using an electric calculator can give a student a much-needed feeling of success.

questioned him about his actions. The youngster replied, "I got the answer! I multiplied 236 by 412. This is the first time I have ever gotten an answer."

There are a variety of activities at different levels of difficulty suited for the laboratories. Some activities may be planned for all students, since they are designed to direct a student's thinking toward certain mathematical concepts. However, other investigations may be a matter of selection by the student.

It provides the teacher with opportunities to study the students' habits of work and thought. In the conventional mathematics classroom a student is assigned exercises and problems for the textbook. These exercises and problems are stated in symbolic and technical language. Each problem provides the necessary data, and the student compares his answer with the *one* provided by the textbook or teacher. The concern of the teacher is primarily the correctness of the answers. All too often the student is able to obtain *the* correct answer with only a hazy understanding of the process and without going through many of the steps one must use in solving real problems.

In the mathematics laboratory, on the other hand, the student is confronted with problems conceived in terms of his environment. Data are not provided. Instead, he must devise a plan of attacking the problem, decide what data are relevant to the problem, and then obtain and process these data. If the process of measurement is involved in obtaining data, then he must work with approximate data and expect his predictions or answers to be only approximate.

Watching a student who is confronted with a problem situation is an effective way to gain insight into many facets of his behavior, perceptions, understandings, and potentials.

It provides motivation for students to improve their proficiencies in mathematical skills and concepts. Investigations in the laboratory should help each student appreciate the significance of mathematics in his daily life. Success in these investigations should build some measure of confidence in his ability to think mathematically. Experiences in the laboratory should also point to the need for him to improve his skills in the use of mathematical symbols. He may try hard to improve whatever skills he has been shown a practical use for.

The student who has always been attuned to failure may for the first time experience the excitement of actually learning something in a mathematics class. He will be working at his own pace in an atmosphere in which he knows he is respected for what he has accomplished. He will be working with real objects that he can see and manipulate himself. When he is allowed to work through problems, using his own thought processes rather than someone else's, he may discover ways in which his own talents are superior to those of his more successful classmates.

Selected Exercises

Chapter 1

1. Construct a model for the positive rational numbers by which students can do the following:

   a) Obtain a feeling for what can be represented by a numeral such as $\frac{5}{8}$.
   b) Develop algorithms for addition and subtraction through manipulating and experimenting with the model
   c) Develop algorithms for multiplication and division through manipulating and experimenting with the model

2. Write a set of exercises that would help students to practice the algorithms for addition, subtraction, multiplication, and division that they developed in exercise 1.

3. Develop a paper that emphasizes one of the following:

   a) The similarities between operations on whole numbers and operations on positive rational numbers
   b) The relation between similarity and congruence of figures
   c) The relation between the following pairings:

      (1) Measure of the diameter of a circle to the circumference of a circle
      (2) Distance to rate (time constant)

(3) Cost of a quantity of an item to the amount of the item purchased

4. Study games that have mathematical applications (see the list in Appendix A for some choices) and report on one that might help students become more creative.

5. Three teachers are teaching a unit on the slide rule to their ninth-grade classes. The first teacher believes in emphasizing skills, the second teacher believes in emphasizing the structure of mathematics, and the third teacher believes in providing students with experiences to help them discover principles and understand what they are doing. In what ways might the classroom activities of the three teachers differ?

6. Describe some procedures you can use to help your class understand that the area of a region is the number of unit squares into which it can be divided. Further describe how students can relate this to the writing of formulas.

7. Show how the manipulation of objects can be related to prime and composite numbers. Then show how this manipulation can be related to the factorization of composite numbers into a product of prime numbers.

8. Tell how you would teach each of the following through the use of manipulative materials:

a) Pi for understanding—so that the students can compute either the circumference or the measure of the diameter of a circle or sphere when the other measure is given, and so that they understand in what situations the circumference might be the easier of the two measures to obtain.

b) Similarity—so that the students understand the concepts of corresponding sides and corresponding angles of similar figures, and so that they discover that corresponding angles are the same size and pairs of corresponding sides are in proportion. How would you react to leaving students with an open-ended question like "When are circles, ellipses, and so on similar?"

9. Make a chart relating classroom activities to the six items listed under "Types of Activities Needed for the Mathematics Classroom." List at least three classroom activities under each of the six items. Write a short statement indicating the relation of the activity to the item. Compare your chart and statements with someone else's. Discuss any differences of opinion.

Chapter 2

1. Describe how you would train students to operate any technical equipment they use in the laboratory.

2. Devise a system that will train students to assemble and work with the materials necessary to do an activity. Include cleaning up after the activity is completed.

3. List several plans of attack for some problem (for example, finding the number of beans in a jar that has an odd shape; finding the height of a tall building or a tall tree; finding the distance between two points in a large field; describing the average person in a classroom). Discuss the relative merits of each plan.

4. You are planning a teacher demonstration for the question "How do you find the volume of a rock?"

   a) How would you guide students to develop a plan of attack?
   b) How would you involve students in the activity?

5. Some of your students are going to investigate the ratio (C:D) for circular objects (C represents circumference and D represents the measure of the diameter), and others are going to investigate similarity using enlargements and reductions of common things (photos, cutouts of similar polygons, and cutouts of similar simple closed curves, for example).

   a) How can you help them identify the problem?
   b) What are several possible plans of attack they might choose?
   c) How would you have them organize their data?
   d) What manner of follow-up activities would you plan?

6. Your classroom consists of very diverse youngsters. There are five separate groups, each learning to do one of the following:

   1. Addition of two three-digit numbers with carrying
   2. Subtraction of two three-digit numbers with borrowing
   3. Multiplication of two two-digit numbers
   4. Factoring two- and three-digit numbers into products of prime numbers
   5. Basic multiplication facts

Available for student use are replicas of 10×10 arrays; objects grouped by ones, tens, hundreds, and thousands; sets of disks; sets of squares; and a balance scale with weights. For each group of students, consider the following questions:

   a) What equipment are they to use?
   b) How can they be led to formulate the problem?

c) What plan of attack would they find most productive?

d) What kind of guidance and follow-up would you provide?

7.   You are a seventh-grade mathematics teacher trying to implement an individualized laboratory approach. You have obtained the assistance of good mathematics people from a high school, a nearby college, or the community (intelligent housewives are often excellent assistants), so that you have one helper for some periods and two for others. How will you use them to best advantage?

Chapter 3

1.   Prepare a guidesheet similar to P-1, Pentominoes, by means of which students are to find all possible arrangements of four cubes. Each cube has to have face-to-face contact with at least one other.

2.   Make an isometric geoboard (see teacher's guide for G-1, The Geoboard); then prepare a guidesheet for students to use in arriving at the formula $\square = \triangle - 2 + 2\bigcirc$. ($\square$, $\triangle$, and $\bigcirc$ represent the same things they represented in guidesheet G-1, The Geoboard.)

3.   What objectives would you have for an activity like measuring distances on the globe? Write a guidesheet for such an activity.

4.   You have two different objects which can land in any one of three ways when thrown (e.g., a cylinder or a frustum of a cone). Assuming that the probabilities for falling any one of the three ways are different for the two objects, a game for identifying the object from knowledge of how it fell on 15 or 20 throws can be played. Write a guidesheet for such a game.

5.   Write a paragraph or two about how an investigation of the question "How many drops of water are in the classroom aquarium?" can be made a challenging and exciting experience. Indicate a plan of attack.

6.   Have a brainstorming session with a group of students (students that you now teach, or of the age that you intend to teach). Use the session to gather a list of questions that they would like your class to investigate.

7.   Repeat exercise 6 with a group of teachers. Specify student age group and, if the group wishes, the mathematical area of concern.

8.   Collect a set of books that contain mathematical puzzles, games, and problems. Using this set of books, put together a list of questions that you would like to have your class investigate.

9. Write a set of objectives (in behavioral or conceptual terms) that you would expect students to attain from studying any two (other than LM-1) of the guidesheets presented in chapter 3.

Chapter 4

1. What changes would you make in your present situation, or in a traditional classroom situation, in order to use the laboratory approach for a weekly change-of-pace day? What type of activities would you plan? What materials would you need?

2. Assume that you and your mathematics colleagues have decided to set up a mathematics laboratory. You intend to have one of the teachers run the laboratory but to allow all to send students to it.

   a) How would you organize this situation to provide for maximum effectiveness of the laboratory teacher?
   b) What reference books would you want in the laboratory?
   c) What other materials would you want in the laboratory?
   d) What are the advantages and disadvantages of setting up such a laboratory?

3. Sketch a floor plan for converting an existing classroom to a laboratory similar to the one you described in exercise 2.

4. Assume that you intend to use the laboratory approach in your classroom; that you have been given a $1000 budget for supplies and equipment; that a shop teacher in the district has agreed to help students construct equipment and other objects if you provide the necessary raw materials; and that the classroom now contains tables, chairs, ample storage space, an overhead projector, and a screen. List each of the following:

   a) The reference books you would buy
   b) The equipment and other objects you would buy
   c) The raw materials you would buy for the shop teacher, and the objects and equipment you would have students make with these materials
   d) Some activities in which you would use some of the objects and equipment
   e) The steps you would take (with materials, classroom layout, class organization, and so on) to maintain a tolerable noise level

5. Sketch a floor plan of the laboratory (see exercise 4) as you would set it up. Show the approximate positioning of tables, chairs, and storage areas. Indicate where you would place materials, equipment, books, and so on.

Chapter 5

1. Evaluate a test that you have recently written with a short paragraph covering each of the following:

   a) What learning objectives are evaluated?
   b) How could the test be improved?
   c) How would you use the test as a diagnostic instrument?

2. Pick one of the broader questions from the test you used in exercise 1. Write a diagnostic test designed to pinpoint deficiencies in the area of coverage. Then write a set of exercises that will help a student remedy any deficiencies indicated by the test.

3. Study the attitude inventory in this chapter (see fig. 5-2) as well as several that have been prepared commercially. Prepare a critique of each; then, on the basis of the critiques, write one that would be more useful to you. If possible, try it out to evaluate the program and yourself.

4. Using an observation form (see fig. 5-4, p. 143), observe and record the progress of a group of students working in a laboratory situation. Write a report appraising the effect of the laboratory situation on student progress.

5. Construct a system for assessing student development and reporting to parents.

6. Find out how your colleagues assess materials of instruction. Using whatever seems valuable, construct a system for a comprehensive evaluation of such materials. Use your system to evaluate the materials you are currently using.

7. Make a list of various means of assessing the teacher and the program. Write a report showing how these materials and procedures could be used to assess the effectiveness of you and your programs. Evaluate yourself and indicate how you could improve your effectiveness as a teacher.

Chapter 6

1. Answer each of the following questions regarding *low achievers*:

   a) What generalizations could you make about the behavior, beliefs, and abilities of this group?
   b) What characteristics should textbooks and other reading materials for this group contain?

c) What characteristics should a learning experience designed for this group encompass?

2.   What are some ways that a teacher can *enhance the self-image* of a low achiever?

3.   List ways in which the atmosphere of a mathematics classroom can stimulate curiosity and participation.

4.   A friend of one of the authors once made the following statement regarding classroom behavior of students: "They will give you what you expect and accept." Write an argument for or against this statement to debate with your colleagues.

5.   You are going to help low achievers in the junior high school learn one of the following topics: volume, place value for whole numbers, computation with decimal fractions, linear equations. Answer each of the following questions about the topic you have chosen:

a) What kinds of experiences with objects would you specify?
b) What procedure would you use to help them understand the topic?
c) Could their concept of estimation be used to help them learn?
d) What are some questions you might use on a diagnostic test for the topic?

6.   In magazine and newspaper articles in 1970 the virtues of a new program for low achievers in the elementary school were extolled. Some of the assumptions of the program are the following:

1. Low achievers need status-building activities.
2. A wise selection of difficult mathematics content should be used with low achievers.
3. Graduate mathematicians should give at least part of the program for low achievers.

a) Consider these assumptions and react to them.
b) Describe areas of mathematics that might be of interest to low achievers but a challenge to high achievers at the same age level.

Appendix B

Use any answers you have written for exercises for chapters 1–6 to help you answer the following:

1.   Pick a topic and write objectives (in behavioral and conceptual terms) for the study of the topic.

2. Write a test that will assess the possession of the knowledge and skills prerequisite to the study of the topic chosen in exercise 1.

3. Write a test that will help determine weaknesses in the attainment of the objectives you wrote for the topic chosen in exercise 1.

4. Construct diagnostic tests related to the test you constructed for exercise 3, that will help you pinpoint difficulties with the attainment of one or two of the objectives you wrote for exercise 1.

5. Write guidesheets like those in this book to help your students attain the objectives you wrote for exercise 1.

6. Present plans for assessing the students' development, the program, the materials, and yourself. Present a plan for reporting what you assess also.

7. Create a teacher's guide to go with the guidesheets written for exercise 5.

a

Appendix A

Following is a list of materials suitable for use in mathematics laboratories, grades 5 through 9. It is not intended as a list of every worthwhile piece of equipment, book, or game that can be used in a mathematics laboratory. The listing of a brand name does not imply that this is the best or only version. Nor must a good mathematics laboratory contain all these items. Rather, they are materials the authors know to be of value. A laboratory containing all of them would be very well equipped and such a laboratory would be a standard to keep in mind when planning and stocking a new laboratory.

Miscellaneous items:

waxed paper; string; marbles; wire; dowel rods of various sizes; wooden beads; felt; golf tees; ice-cream sticks; pipe cleaners; play money; mail-order catalogs; filament tape; strips of wood (1 in. × 1½ in. × 10 ft.); sheets of ½-in. plywood; pegboard; corrugated cardboard; colored chalk; balsa wood; Tri-Wall® cardboard (Tri-Wall Containers, Inc., 1 Dupont St., Plainview, N.Y. 11803); rubber balls (5 in. to 8 in. in diameter and of uniform color); colored yarn; plastic straws; elastic thread; Con-tact® paper; spray paint; paintbrushes.

Office supplies:

pencils; felt-tip pens (with both permanent and water-soluble colored ink); masking tape; brightly colored plastic tape; transparent tape; rubber cement; stapler and staples; paper clips; 5×8-in. index cards;

graph paper; thumb tacks; map tacks; spirit masters; large paper cutter; acetate sheets; rubber bands; poster boards and blotters of various colors; sheets of ⅛-in. clear plastic.

Instruments for measuring and counting:

50-foot tape measure; rulers; protractors; magnetic compasses; thermometers; graduated cylinders; levels; stopwatches; hand counters; odometers; vernier calipers; micrometer calipers; distance wheels for maps; meter sticks; printed scales (tenths of inches, hundredths of feet).

Drawing instruments:

drawing boards; T squares; triangles; drafting set; Doric lettering set; compass with extension arm; protractors; proportional dividers; small clear plastic drawing board and T square for overhead projector.

Display and presentation devices:

chalkboard; overhead projector and screen; transparency pens; graph board; templates for chalkboard use (parallel lines, perpendicular lines, protractor, 30°-60°-90° triangle); corkboard; acetate envelope.

Mathematics games and puzzles:

Quinto®, Sum-Times®, Awari®, Twixt® (Minnesota Mining & Manufacturing Co., P.O. Box 3717, St. Paul, Minn. 55101); *Numo, Ranko* (Midwest Publications Co., Inc., P.O. Box 307, Birmingham, Mich. 48008); *Radix®* (J. W. Lang, Box 224, Mound, Minn. 55364); *Even-Steven, Hi-Q, Pythagoras* (Kohner Bros., Inc., 1 Paul Kohner Pl., East Paterson, N.J. 07407); *Equations* (Science Research Associates, Inc., 259 E. Erie St., Chicago, Ill. 60611); *Numble™* (Selchow & Righter Co., Bay Shore, N.Y. 11706); *Soma®* (Parker Bros., Inc., Salem, Mass. 01970); *Tower of Hanoi,* by 3-World's Co. (Marshall Field & Co., distributor, 111 N. State St., Chicago, Ill. 60603); *Score Four™* (Funtastic, 5902 Farrington Ave., Alexandria, Va. 22304); *Hex* (J. W. Spear & Sons, Ltd., Enfield, Middlesex, England); *Wff 'n Proof®* (Wff 'n Proof, Ann Arbor, Mich.); *Mem®* (Stelledar, Inc., 1700 Walnut St., Philadelphia, Pa. 19103); *Trimino™, Pitfall™* (Creative Playthings, Inc., distributor, Princeton, N.J. 08540); *3-D Tic-Tac-Toe®* (Crestline Manufacturing Co., 603 G E. Alton St., Santa Ana, Calif. 92705).

Geometric models:

colored pieces of cardboard or wood pieces of different geometric shapes; models of common solids; plastic model showing sections of a cone; map projecton globe; solids of revolution; cone-sphere-cylinder demonstration model of volume; mirrors for study of symmetry; sheets of clear plastic (on which geometric figures can be drawn); viewbox for showing the three orthographic views of objects; box of 1000 1-in. cubes; pictures of objects that have been sectioned; pictures of geometric shapes in everyday life; geoboards; colored rubber bands; bundles of sticks that can be pegged together (one 7-in., one 13-in., four 8-in., and two 11-in. sticks in each).

Equipment for fieldwork:

tapes (50 ft. or 100 ft.); taping pins (30-D nails punched through pieces of red or orange cardboard); plane table (drawing board); tripod (or tall wastebasket on a three-legged stool); alidade; transit; sextant; clinometer; plumb bobs; bicycle wheel for measuring distance; wooden pegs.

Devices for linear and area measure:

meter sticks; yardsticks; a set of congruent sticks ($\frac{3}{16}$-in. dowel rods) in three colors; adding machine tape; meter wheel (a plywood wheel 1 meter in circumference mounted on form for rolling); demonstration vernier; geoboard; squared paper (1 × 1-in. grid); square-inch grid on clear acetate sheets; square-foot pieces of floor tile; globe of the earth; blocks of wood and cylinders to measure.

Materials for numeration and computation:

electric calculators; slide rules; abacuses; nomographs; binary adders; objects grouped with respect to decimal and nondecimal bases.

Items for demonstrating probability:

200 marbles of three colors for sampling experiments; dice; large paper clips; deck of playing cards; wooden block with four circular wells for coin tossing; three 1-in. cubes with spots for dice (one loaded); wooden cylinders (to show ratio of frequencies of three events); frustum of cone (toothpaste caps).

Equipment for demonstrating similarity:

set of similar figures (triangles, quadrilaterals, drawings, photographs, postage stamps); sets of similar objects (spheres, cylinders, cubes, boxes, model cars); protractors; pantograph; rubber bands (⅛ in. wide and 3½ in. to 4 in. long); transparent acetate grids (1 in. × 1 in., ½ in. × ½ in., ¼ in. × ¼ in.); grid paper (1 in. × 1 in., 2 in. × 2 in., 1 in. × 2 in.); mirrors (plane and curved); road maps; house plans.

Objects that show ratio:

ratio compass; egg beater; bicycle with three gears; pi circle (made of ½ in. plywood with a diameter of 1.00 ft.); pulleys; block and tackle; wood and plastic foam blocks of same size for a study of density; oatmeal box for making a model of a car cylinder; box of objects for comparison (such as colored blocks, nails, hinges and screws, bolts, and washers); set of sticks with lengths in the extended ratio (3 : 4 : 5).

Materials to show rational numbers:

template for drawing circles on chalkboard and marking off congruent sectors; elastic percent indicator; yardsticks covered with strips of paper marked to show decimal subdivisions; template for drawing a square foot on chalkboard and marking 3, 4, 6, or 12 congruent parts.

Materials to show volume:

1000 1-in. cubes; a cubic-foot model; 3 faces of a cubic-yard model (see fig. 1); containers for liquid measure (cup, pint, quart, gallon); medicine dropper; pipettes; graduated cylinders (1000 c.c.); tin cans; 1 quart of rice; cone, cylinder, pyramid, and prism models; wooden blocks.

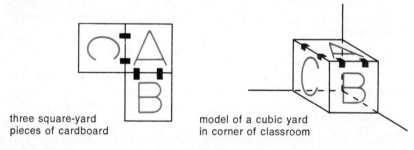

three square-yard   model of a cubic yard
pieces of cardboard   in corner of classroom

Fig. 1

Equipment for demonstrating weight:

homemade balances; large paper clips and fishing sinkers (1 oz. and 2 oz.) for units of weight; postage scales (in ounces to 5 or 10 pounds); hand spring scales (grams, ounces); objects to weigh (blocks of wood, plastic foam).

Tools:

pliers; hammers; screwdrivers; screws; nails; bench vises; carpenter's squares; wood glue; sandpaper; knives; hand drills; stapler with ½-inch staples; wire cutters; wood chisels; hacksaws.

Reference Books and Learning Kits

Abbott, Edwin. *Flatland.* New York: Barnes & Noble, 1963.

Adler, Irving. *The Giant Golden Book of Mathematics.* New York: Golden Press, 1960.

————. *Logic for Beginners through Games, Jokes and Puzzles.* New York: Day, 1964.

Asimov, Isaac. *An Easy Introduction to the Slide Rule.* Boston: Houghton Mifflin, 1965.

————. *Realm of Numbers.* Boston: Houghton Mifflin, 1959.

Babcock, David R. "Cardboard Carpentry." *Instructor* 79 (November 1969): 99–102.

Barr, Steven. *Experiments in Topology.* New York: Thomas Y. Crowell, 1964.

Behavioral Research Laboratories. Reading, Mass.: Addison-Wesley, 1963. A series of programed texts:

Addition
Basic Mathematics: A Problem Solving Approach
Division
Multiplication
Subtraction

Bendick, Jeanne. *How Much and How Many: The Story of Weights and Measures.* New York: McGraw-Hill, 1947.

Bendick, Jeanne, and Levin, Marcia. *Mathematics Illustrated Dictionary.* New York: McGraw-Hill, 1965.

————. *Take a Number: New Ideas Plus Imagination Equals More Fun.* New York: McGraw-Hill, 1961.

Bergamini, David, et al. *Mathematics.* New York: Time Inc., 1963.

Bowers, Henry and Joan F. *Arithmetical Excursions: An Enrichment of Elementary Mathematics.* New York: Dover, 1961.

Brandes, Louis. *Cross-Number Puzzles.* Portland, Maine: Walch, 1957.

————. *4 The Math Wizard.* Portland, Maine: Walch, 1962.

————. *Yes, Math Can Be Fun!* Portland, Maine: Walch, 1960.

Cross-Number Puzzles. Chicago: Science Research Associates, 1966–70. A series of 4 kits:

Murfin, Mark, and Bazelon, Jack. *Decimals and Percent*
————. *Fractions*
————. *Whole Numbers*
Story Problems

Degrazia, Joseph. *Math Is Fun.* New York: Emerson, 1954.

Diggins, Julia. *String, Straightedge, and Shadow: The Story of Geometry.* New York: Viking, 1965.

Dorrie, Heinrich. *100 Great Problems of Elementary Mathematics.* New York: Dover, 1965.

Earl, Boyd; Hauck, William; McFadden, Myra; and Reigh, Mildred. *Bucknell Mathematics Self-Study System.* New York: McGraw-Hill, 1965, 1967. Three sets of programed materials for self-study.

Fitzgerald, William, et al. *Laboratory Manual for Elementary Mathematics.* Boston: Prindle, Weber & Schmidt, 1969.

Friskey, Margaret, ed. *About Measurement.* Chicago: Melmont, 1965.

Gardner, Martin. *Mathematical Puzzles.* New York: Thomas Y. Crowell, 1961.

_____. "Of Sprouts and Brussels Sprouts: Games with a Topological Flavor." *Scientific American* 217 (July 1967):112–15.

Golomb, Soloman W. *Polyominoes.* New York: Scribner, 1965.

Herrick, Marian C. *Modern Mathematics for Achievement.* Boston: Houghton Mifflin, 1966. A series of 8 booklets for students of low ability.

Highland, Harold. *The How and Why Wonder Book of Mathematics.* New York: Wonder Books, 1961.

Hogben, Lancelot. *Wonderful World of Mathematics.* Garden City, N.Y.: Doubleday, 1968.

Hunter, James. *More Fun with Figures.* New York: Dover, 1966.

Johnson, Donovan. *Games for Learning Mathematics.* Portland, Maine: Walch, 1960.

_____. *Paper Folding for the Mathematics Class.* Washington: National Council of Teachers of Mathematics, 1957.

Johnson, Donovan, et al. "Exploring Mathematics on Your Own." New York: McGraw-Hill, 1960–63. A series of 18 books:

Adventures in Graphing	*Number Patterns*
Basic Concepts of Vectors	*Probability and Chance*
Computing Devices	*Pythagorean Theorem*
Curves in Space	*Sets, Sentences, and Operations*
Finite Mathematical Systems	*Shortcuts in Computing*
Fun with Mathematics	*Topology*
Geometric Constructions	*Understanding Numeration Systems*
Invitation to Mathematics	*World of Measurement*
Logic and Reasoning	*World of Statistics*

Kaplan, Philip. *Posers: Eighty Delightful Hurdles for Reasonably Agile Minds.* New York: Harper & Row, 1963.

Larsen, Harold D., et al. *Enrichment Program for Arithmetic.* New York: Harper & Row, 1956–60. A series of booklets on topics for grades 4–8.

Lewis, K., and Pearcy, J. F. F. *Experiments in Mathematics.* Stages 1, 2, and 3. Boston: Houghton Mifflin, 1967.

May, Lola J. *Elementary Mathematics: Enrichment.* New York: Harcourt, Brace & World, 1966. Four paperback booklets.

Meyer, Jerome. *Fun with Mathematics.* New York: World Publishing, 1952.

Mueller, Francis, and Hach, Alice. *Mathematics Enrichment.* Programs D and E. New York: Harcourt, Brace & World, 1963.

National Council of Teachers of Mathematics. *Enrichment Mathematics for the Grades.* Twenty-eighth Yearbook. Washington: the Council, 1963.

_____. "Experiences in Mathematical Discovery." Washington: the Council, 1966. A series of 5 volumes:

1: *Formulas, Graphs, and Patterns*
2: *Properties of Operations with Numbers*
3: *Mathematical Sentences*
4: *Geometry*
5: *Arrangements and Selections*

_____. "Topics in Mathematics for Elementary School Teacher." Second Series. Washington: the Council, 1969. A series of 6 books:

Collecting, Organizing and Interpreting Data
Graphs, Relations, and Functions
Informal Geometry
Logic
Measurement
Symmetry, Congruence, and Similarity

Niven, Ivan, and Zuckerman, H. S. "Lattice Points and Polygonal Area." *American Mathematical Monthly* 74 (December 1967):1195–1200.

Nuffield Mathematics Project. New York: Wiley, 1967, 1968.

Beginnings
Computation and Structure (2 vols.)
Desk Calculators
How to Build a Pond
I Do, and I Understand
Mathematics Begins
Pictorial Representation
Shape and Size (2 vols.)
Your Child and Mathematics, by W. H. Cockroft

Northrop, Eugene. *Riddles in Mathematics, A Book of Paradoxes.* Princeton, N.J.: Van Nostrand, 1944.

Peck, Lyman C. *Secret Codes, Remainder Arithmetic, and Matrices.* Washington: National Council of Teachers of Mathematics, 1961.

Polya, George. *Mathematical Discovery: On Understanding, Learning, and Teaching Problem Solving.* Vols. I and II. New York: Wiley, 1962.

Proctor, Charles M., and Johnson, Patricia. *Computational Skills Development Kit.* Chicago: Science Research Associates, Inc. 1965.

Proctor, Charles M., and Lacy, Joseph. *Algebra Skills Kit.* Chicago: Science Research Associates, Inc. 1969.

Rapp, Dale R. *Arithmetic Fact Kit.* Chicago: Science Research Associates, Inc. 1969.

Ravielli, Anthony. *An Adventure in Geometry.* New York: Viking, 1957.

Reid, Constance. *From Zero to Infinity.* New York: Thomas Y. Crowell, 1960.

Ringenberg, Lawrence A. *A Portrait of 2.* Washington: National Council of Teachers of Mathematics, 1964.

Simon, Leonard, and Bendick, Jeanne. *The Day the Numbers Disappeared.* New York: McGraw-Hill, 1963.

Smith, David Eugene. *Number Stories of Long Ago.* Washington: National Council of Teachers of Mathematics, 1919.

Strater, William W. *Five Little Stories.* Washington: National Council of Teachers of Mathematics, 1960.

Understanding Modern Mathematics. New York: Macmillan, 1963. A series of programed texts:

Bases and Numerals
Clear Thinking
Factors and Primes
Modular Systems
Number Sentences
Points, Lines, and Planes
What Are the Chances?

Vergara, William. *Mathematics in Everyday Things.* New York: Harper & Row, 1959.

Vorwald, Alan, and Clark, Frank. *Computers: From Sand Table to Electronic Brain.* New York: McGraw-Hill, 1961.

Wenninger, Magnus J. *Polyhedron Models for the Classroom.* Washington: National Council of Teachers of Mathematics, 1966.

Willerding, Margaret. *Mathematical Concepts: A Historical Approach.* Vol. 5. Boston: Prindle, Weber & Schmidt, 1967.

Wirtz, Robert; Botel, Horton; and Sawyer, W. W. *Math Workshop.* Levels C, D, E, and F. Chicago: Encyclopaedia Britannica, 1966.

b

Appendix B
Sample Investigations: Ratio

The concept of ratio is a useful tool in the development of other topics in mathematics. It is used, for instance, in geometry to discover the relation between the circumference and diameter of a circle; in analytic geometry to represent the slope of a line; and in other junior high and secondary school courses to indicate various relationships. Number sentences about equal ratios (proportions) can be used to motivate the study of equations. A few other applications of ratio are earned-run average; batting average; compression in an automobile engine; gasoline mileage for an automobile; earning power of stocks; population density; musical chords; specific gravity; relative humidity; strength of solution (chemistry); conversion of measurement units; similarity (geometry); and probability.

Some students study ratio in the sixth grade, but many others are not introduced to this topic until high school. Those who do not study ratio in elementary or junior high school are being denied something that could be a valuable aid to them in learning other topics in mathematics.

The authors feel strongly that all students should be introduced to this concept as early as possible, and that the laboratory method is the best way of presenting it. For that reason they have chosen to give a laboratory development of this topic. It is intended to serve as a useful example of implementation of the laboratory approach.

To implement the study of ratio through the laboratory approach, the teacher should specify performance objectives, decide which activities will promote attainment of the specific objectives, write guide-sheets for these activities, and collect the necessary materials. If the teacher has imbued his students with the concept of active learning, helped them develop responsibility for much of their own learning, arranged the classroom to facilitate individualized active learning, and adopted procedures for evaluating individual student progress, then he is ready to start developing a concept such as ratio in the manner outlined by the guidesheets in this appendix.

If the teacher has not done these things, then much remains to be done. First, he must get his students accustomed to moving about, manipulating materials, and collecting data. Second, he must teach the students the new roles of teacher and student in the laboratory (see chapter 2). Third, he must get students used to working in small groups or as individuals in different places in the room and using different types of equipment. Finally, he must adopt a procedure for evaluation that will enable him to measure each student's development (see chapter 5).

Overview of the Study of Ratio

Writing performance objectives for the study of ratio involves the following:
1. Developing an overview of what ratio is, and indicating the symbolism, relevant vocabulary, principles, skills, and types of applications to emphasize
2. Determining what knowledge is prerequisite for the study of ratio
3. Determining performance objectives

We will undertake the first task here. This development is in line with the authors' concept of ratio and how it should be studied. It is intended as background for teachers so that they can understand better why the authors have developed the guidesheets as they have. This section is not intended for students.

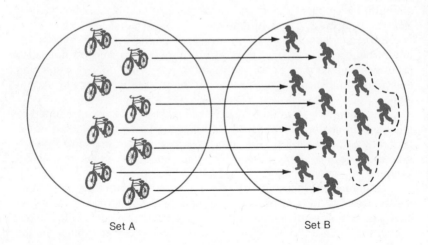

Set A   Set B

Comparisons between the number of elements in two sets can be made in two ways. The first involves subtraction and seeks to answer such questions as: Are there enough? Does one set have more elements than the other? How many more are needed? This method involves the one-to-one matching of the elements of one set (set A) with elements of a subset of another set (set B), where set B has at least as many elements as set A. The cardinality (number of elements) of the set of elements left unmatched in set B represents how many more elements are in set B than in set A. For example, if eight bicycles are matched in this way with twelve boys, we would find that there are four more boys than bicycles (see the figure on the preceding page).

The second method of comparing the number of elements in two sets involves matching the cardinalities of the sets. For example, if the cardinality of the set of eight bicycles is matched with the cardinality of the set of twelve boys, we say that the *ratio* of the number of bicycles to the number of boys is 8 to 12, and we write this as (8:12). If we partition each set into four equivalent subsets, we see that there are two bicycles for every three boys. Therefore, the ratio (2:3) can also be used to compare the number of bicycles with the number of boys. In either type of comparison mentioned above, the objects in the sets need not be of the same kind, since the comparison is between cardinalities.

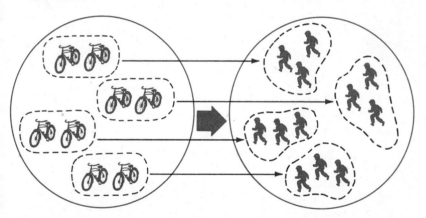

Assume that the cardinalities of two sets are compared in a given direction, and assume that the two sets are then partitioned into the same number of equivalent subsets. If the cardinalities of a pair of subsets—one from the first set and the other from the second—are compared in the same direction, the two ratios thus formed are said to be *equal*. In the example just given, (8:12) is equal to (2:3).

Ratios can also be used to compare two measurements, since a measurement can be thought of as a count of the number of units of measurement. In this way we can obtain ratios with fractions, such as $(\frac{1}{2} : \frac{2}{3})$, or with decimals, such as (2.50:5.37).

Ratios are not being considered as numbers on which operations are defined; therefore we use the form (3:4) rather than $\frac{3}{4}$. This form also makes it easier to generalize the concept of ratio. The numbers that make up the ratio are called the *terms* of the ratio.

It is possible to compare the cardinality of more than two sets at the same time. For example, a comparison like (4:7:5) might be used to show the available plant foods (nitrates, phosphoric acid, and potash) in a fertilizer. The comparison (2:3:4) represents the relative frequencies of middle C, G, and high C on a piano. (1:2:3) represents the relative amounts of cement, sand, and gravel in a good mixture of cement. The Egyptians used (3:4:5) as the relative lengths of the sides of a rope triangle with which to obtain square corners in surveying. A comparison with three or more terms will be called an *extended ratio*.

The following diagram shows several ways of comparing the cardinalities of sets A, B, C, and D.

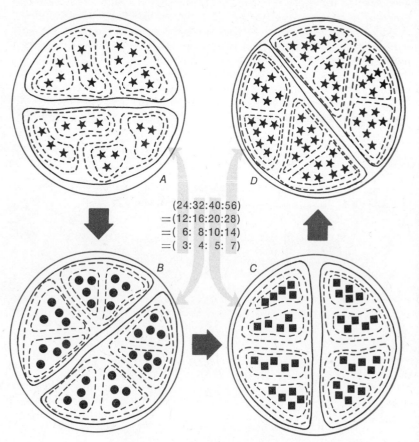

$(24:32:40:56)$
$=(12:16:20:28)$
$=(\ 6:\ 8:10:14)$
$=(\ 3:\ 4:\ 5:\ 7)$

Fig. B-1

There are certain aspects of ratio that must be learned in order to make full use of this concept. These include *equality* of ratios and extended ratios expressed by proportions and proportionalities, and the *product rule.*

Equality.   In a previous example we found that $(8:12) = (2:3)$, since each ratio was used to compare the number of bicycles with the number of boys. Concrete made with the relative amounts of cement, sand, and gravel $(1:2:3)$ would have the same characteristics as that made by the formula $(4:8:12)$; therefore $(1:2:3) = (4:8:12)$. Figure B-1 illustrates why $(24:32:40:56)$, $(12:16:20:28)$, $(6:8:10:14)$, and $(3:4:5:7)$ are equal. Essentially, then, ratios or extended ratios that can be used to compare the same sets or the same measures, as just indicated, are equal.

Definition: Given four nonnegative real numbers a, b, c, and d, $(a:b) = (c:d)$ if and only if there exists a positive real number k such that $c = ka$ and $d = kb$. This definition can easily be generalized to include equality of extended ratios. For any integer $n \geq 2$ and non-negative real numbers $a_1, \ldots, a_n, b_1, \ldots, b_n$, $(a_1: a_2: a_3: \ldots a_n) = (b_1: b_2: b_3: \ldots b_n)$ if and only if there exists a positive real number k such that $b_1 = ka_1$, $b_2 = ka_2$, $b_3 = ka_3$, $\ldots$, $b_n = ka_n$.

In each case the number k mentioned in the definition above is called the *constant of proportionality.* A sentence that states that two ratios are equal is a *proportion;* a sentence that states that two extended ratios are equal is a *proportionality.*

In some texts the symbol $::$ is used instead of $=$ to indicate that two ratios or extended ratios are equal. Both $::$ and $=$ in a proportion or proportionality are read "as." For example, $(1:2) = (3:6)$ is read "1 is to 2 as 3 is to 6," and $(3:4:5) = (9:12:15)$ is read "3 is to 4 is to 5 as 9 is to 12 is to 15."

Product Rule for a Proportion.   In a proportion the end terms (8 and 3 in the following example) are called the *extremes,* and the middle terms (12 and 2 in the example) are called the *means.*

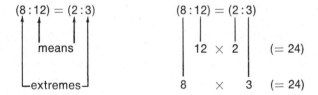

In this example the product of the extremes is equal to the product of the means $(8 \times 3 = 12 \times 2)$. For all true proportions these two products are equal. A simple proof based on the definition of equality of two ratios follows.

If $(a:b) = (c:d)$, then there is a positive real number k such that $c = ka$ and $d = kb$.

Therefore, by the associative and commutative properties of real numbers, we have $ad = k(ab) = bc$ or $ad = bc$.

It should be evident that the student should know a number of things before he can successfully study ratio in the laboratory. Students with a background only in counting numbers should be able to (a) operate with counting numbers; (b) recognize and understand the equal sign; (c) solve simple linear equations; (d) recognize when such equations are true or false; (e) measure length, using the British and metric systems, and operate with these measures; and (f) partition sets into equivalent subsets. Students should know what line segments are and how to tell whether they are congruent. If the terms are to include decimal and common fractions, students should be able to do the preceding when such numbers are involved.

Finally, the teacher must be able to restate an overview of the topic as performance objectives. By having such explicit statements of performance, the teacher is better able to determine when activities will help the students attain the objectives. The next section develops an overview of ratio—as the authors would teach it—in terms of expected performance at the termination of study.

Performance Objectives for Experiments in Ratio

The list of performance objectives below is not meant to be exhaustive. It is given only to show how to state performance objectives for obtaining a level of performance commensurate with the overview. The letters can represent whole numbers or nonnegative real numbers, depending on the students.

The student should be able to do the following:

1. Write a ratio in the form $(a:b)$ to compare—
   a) the cardinalities of two given sets of objects;
   b) the lengths of two given line segments.
2. Write an extended ratio in the form $(a:b:c)$ to compare—
   a) the cardinalities of three sets;
   b) the lengths of three line segments.
3. Show—
   a) two sets of objects whose cardinalities can be compared by a given ratio;
   b) two line segments whose lengths can be compared by a given ratio.
4. Show—
   a) three sets of objects whose cardinalities can be compared by a given extended ratio;
   b) three line segments whose lengths can be compared by a given extended ratio.
5. Given two sets of objects whose cardinalities can be compared by a given ratio, partition each set into a specified number of equivalent subsets and write a ratio to compare the cardinalities of two such subsets.

6. Given two line segments whose lengths can be compared by a given ratio—

    a) divide each into a specified number of congruent pieces and write a ratio to compare the lengths of two such pieces;

    b) construct line segments twice the length of the originals and write a ratio to compare the lengths of two such segments.

7. Write a ratio or an extended ratio to represent such statements as the following:

    a) There are three saddles for four horses.

    b) For every three desks there are four students.

    c) For every three fine restaurants there are four hamburger drive-ins and five fried-chicken houses.

8. Change a proportion such as $(a:b) = (c:d)$ into product form $a \times d = b \times c$.

9. Given a, b, and c, find a replacement for $\square$ that will make the sentence true:

    a) $(a:b) = (c:\square)$ *c)* $(a:b) = (\square:1)$

    b) $(a:b) = (\square:c)$ *d)* $(a:b) = (\square:100)$

10. Write and solve a proportion for situations such as the following:

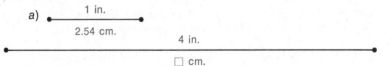

    b) If a bicycle wheel makes 10 turns in moving 74.2 feet, how far will it move in making 31.45 turns?

Descriptions of Selected Laboratory Activities

Ratio and the related vocabulary can be introduced as early as the middle elementary school grades. There are, however, many uses for ratio in the secondary schools—and many techniques for handling them that are appropriate only to students at that level. Therefore it seems advisable to have activities for at least two levels. Fourth- and fifth-grade students are familiar with the idea of matching elements of one set with elements of a second set, and a ratio merely represents a comparison of the cardinalities of two sets. Matching and comparing activities can therefore be the basis for an introduction to ratio. The symbolism and vocabulary are simple enough for such students. A matching activity can take the form of having students write $(2:3)$ to represent two objects of the first set for every three objects of the second set, or two units of measure for the first segment to three units of measure for the second. This interpretation makes it reasonable to partition two sets into the same number of equivalent subsets and to cut two segments into congruent pieces, thereby developing the ideas of equal ratios (proportions). Fourth- and fifth-grade students should also be able to reverse the process and construct sets and segments whose comparisons match given ratios.

Problem situations should also be presented. They should provide students with an opportunity to use equal ratios to make predictions. In such problem situations it may be helpful to have students describe the sets or measures being compared: for example, (number of paper clips : number of lead weights), or (number of turns of the wheel : number of feet the wheel moves).

The following activities are grouped according to difficulty level. Activities R-A-1 through R-A-5 are recommended as introductory activities. The difference between the remaining A activities and the B activities is that the former require only counting numbers whereas the latter require real numbers. The sequence suggested for level A is activities 1, 2, 3, 4, and 5; and any three of activities 6, 7, 8, 9, and 10. Students who have trouble, however, should do some of the activities they skipped and should be given individual attention.

The first five level-A activities should precede the level-B activities. Students doing the level-B activities should do the first five before doing any others. The remainder of the level-B activities should be done by need and choice.

Audio tapes of the guidesheets can be very useful for students with low reading levels. Such students can listen to the tape while reading the guidesheets and manipulating the objects.

The Pretest. Each series of guidesheets should have a pretest to check for attainment of the prerequisite knowledge and skills. The following pretest is an example of one that could be used before work on the B guidesheets on ratio is begun. It has been set up to follow closely the statement of prerequisite knowledge and skills that appeared earlier in this appendix. Some statements of prerequisites might well allow the use of commercially prepared tests.

1. Which of the following means the same as 2.6 feet?
   a. 2 feet 6 inches   c. 2 feet and 6 tenths of a foot
   b. 26 inches   d. 26 feet
2. Which statement is true?
   a. 3.149 to the nearest tenth is 3.2.
   b. 8.36 to the nearest tenth is 8.3.
   c. 7.228 to the nearest hundredth is 7.23.
   d. 17.044 to the nearest hundredth is 17.05.
3. Multiply.
   a. $.5 \times 3 = $ _____   c. $1.51 \times .3 = $ _____
   b. $1.5 \times 3 = $ _____   d. $34.2 \times 1.57 = $ _____
4. Divide.
   a. $160 \div 3.2 = $ _____   c. $1.60 \div 32 = $ _____
   b. $16.0 \div 32 = $ _____   d. $16.0 \div 3.2 = $ _____
5. Solve each equation for $\square$.
   a. $3 \times 4 = 12 \times \square$ c. $3 \times 4 = \square \times 12$
   b. $\square \times 3 = 4 \times 12$ d. $16 \times \square = 23.6 \times 8.5$

6. Which equations are true?

a. $3 \times 12 = 4 \times 9$ c. $3 \times 7 = 4 \times 6$
b. $5 \times 12 = 7 \times 10$ d. $5 \times 14 = 7 \times 10$

7. Find the measure of each segment in inches.

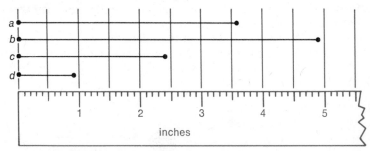

8. Draw line segments of the following lengths.

a. 10 cm.   c. 25.4 cm.
b. 91.4 cm.   d. 12.7 cm.

9. Solve each problem.

a. *A, B,* and *C* are points on a line. *B* is between *A* and *C*. The length of segment *AB* is 18.3 cm. Segment *BC* is 20.0 cm. What is the length of segment *AC* (in cm.)?

b. *E* and *F* are points on a line. The length of segment *EF* is 245 cm. What is this length, expressed in meters?

10. For each of the following line segments, name the line segment on line *PQ* that is congruent to it.

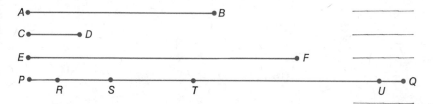

11. Partition the given set in each of the following ways:

a. Two equivalent subsets that do not intersect

b. The two subsets in *a* into three equivalent subsets

c. The six subsets in *b* into five equivalent subsets

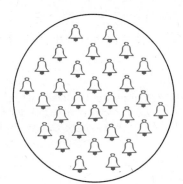

The Posttest.   Each series of guidesheets should be followed by a posttest. This test should be designed to test for attainment of the performance objectives. An example of such a test for the level-A guidesheets on ratio follows. Note that this test has been set up to follow closely the performance objectives for experiments in ratio listed earlier in this appendix.

1. Write a ratio for each of the following:
   a. The comparison of the number of elements in set A with the number of elements in set B

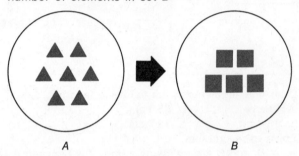

   b. The comparison of the lengths of segments *AB* and *CD*.

2. Write an extended ratio for each of the following:
   a. The comparison of the numbers of elements in sets C, D, and E

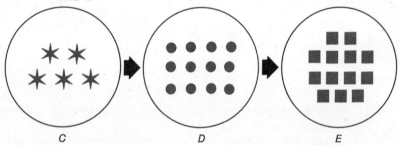

   b. The comparison of the lengths of segments *AB*, *CD*, and *EF*

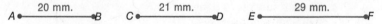

3. Draw each of the following:
   a. Line segment *CD* such that the length of *AB* is to the length of segment *CD* as (2:9).

b. A set of circles so that the ratio of the number of triangles to the number of circles is (7:5).

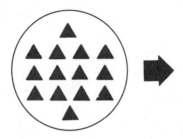

4. Draw each of the following:
   a. Three sets in which the numbers of elements can be compared by the extended ratio (3:4:5)
   b. Three line segments whose lengths can be compared by the extended ratio (2:3:4)

5. Write two different ratios to compare the number of elements in the two sets.

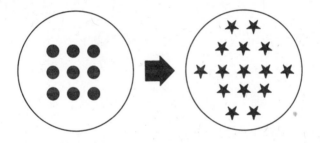

6. a. Draw line segments of lengths 6 centimeters and 12 centimeters. Write a ratio to compare their two lengths (shorter to longer).
   b. Next, divide each line segment into three congruent pieces. Write another ratio to compare the length of one part of the shorter segment with one part of the longer segment.
   c. Write a ratio to compare the lengths of line segments twice as long as those drawn in *a* (shorter to longer).

7. Write a ratio or an extended ratio to represent each of the following:
   a. There were twice as many girls as boys at the dance.
   b. For every two clear streams there is one that is polluted.
   c. In the high school parking lot there were 90 motorbikes and 60 cars belonging to students for every 150 bicycles (also belonging to students).

8. In each problem, write a replacement for ☐ that will make the equation true.

   a. $(3:5) = (☐:10)$ ☐ = _____
   b. $(4:5) = (60:☐)$ ☐ = _____
   c. $(3:5) = (☐:100)$ ☐ = _____
   d. $(35:7) = (☐:1)$ ☐ = _____

9. Solve each problem, using ratios (or extended ratios) and proportions (or proportionalities).

   a. Chris had to read a 320-page book for a test on Monday. She read the first 40 pages in 1 hour and 20 minutes on Sunday afternoon. At this rate, how many minutes would it take her to read the whole book?

   b. Copsan, a robber, stole $560 worth of silver on Friday. He pillaged $1680 worth of famous paintings on Saturday. He purloined $2800 worth of jewels on Sunday. Each week the worth of his Friday, Saturday, and Sunday robberies continues in the same extended ratio. If he steals $2100 worth of goods on the following Saturday, what is the worth of his loot on each of his other two *workdays*?

R-A-1

problem: How do you use ratios to compare the numbers of objects in two sets?

materials:

Envelope containing 15 paper clips, 18 nails, 24 screws, 6 hinges, 9 triangles, 24 squares

procedure:

Example ▶ Refer to figure 1. The two sets are compared element to element. You can see that there are _____ more stars than squares.

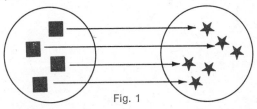

Fig. 1

Refer to figure 2. The two sets are compared set to set. There are _____ squares and _____ stars. We can write the comparison in the form (4:6). We call this comparison a *ratio.*

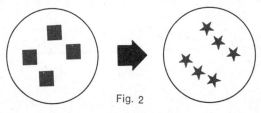

Fig. 2

Refer to figure 3. Each set is separated into two equivalent subsets. They are compared subset to subset. The ratio (4:6) describes the comparison by sets. The ratio (2:_____) describes the comparison by subsets.

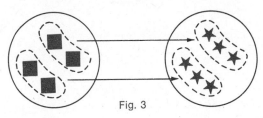

Fig. 3

1. Complete each ratio. Watch the direction of the arrow to see which set comes first.

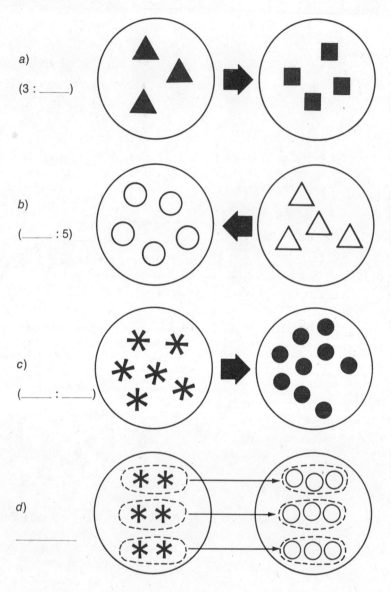

a)

(3 : _____)

b)

(_____ : 5)

c)

(_____ : _____)

d)

_____

2. In exercise 1 we compared (circle one)

    a) the shapes of objects in two sets;
    b) the sizes of objects in two sets;
    c) the colors of objects in two sets;
    d) the numbers of objects in two sets.

3. Write three ratios to compare the number of triangles with the number of squares in three different ways.

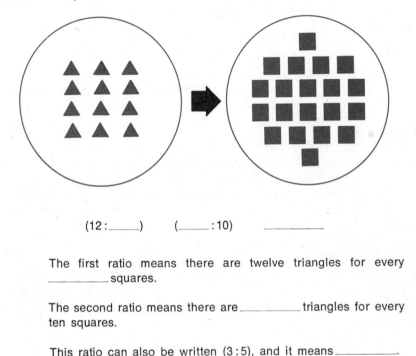

(12 :_____)   (_____: 10)   _____

The first ratio means there are twelve triangles for every _____ squares.

The second ratio means there are_____triangles for every ten squares.

This ratio can also be written (3:5), and it means_____.

4. Refer to the envelopes of paper clips, nails, screws, and hinges. Write a ratio for each of the following:

      a) The number of paper clips to the number of nails_____
      b) The number of screws to the number of hinges_____
      c) The number of squares to the number of triangles_____

5. Here is how one group of students worked exercise 4.

      a) Tom wrote (15:18) to describe the ratio of the number of paper clips to the number of nails.
         Marcia wrote the ratio (5:6).
         Paul wrote (18:15).
         Which were right?_____

      b) David wrote (6:24) to describe the ratio of the number of screws to the number of hinges.
         Dana wrote the ratio (12:3) to describe the same ratio.
         Diane wrote (24 : 6).
         Which were right?_____

      Write another ratio to show this comparison._____

c) Three students wrote (3:8) to show the ratio of the number of squares to the number of triangles. Were they right?

_____

Write two more ratios that show this comparison._____ and_____.

6. Write a ratio to compare the number of boys with the number of girls in your class._____

7. Write a ratio to represent each of the following:

a) There are eighteen bicycles and thirty students._____
b) There are two bicycles for every three students._____
c) There are twice as many students as there are bicycles.

_____

R-A-2

problem: How do you use ratios to compare lengths of objects?

materials:

> 12-inch ruler
> 1 red, 1 green, 1 yellow, and 1 black stick
> Nail
> Clothespin
> cm. ruler

procedure:

Example ▶ Measure the red stick and the green stick in inches. The red stick is _____ inches long. The green stick is _____ inches long.

The ratio (_____ :6) can be used to compare the length of the red stick with that of the green stick. The ratio (1 : _____) could also be used. This means "one for every two." The red stick has a length of 1 inch for every _____ inches of the green stick.

1. Measure each object and record the length.

   a) red stick   _____ in.
   b) green stick   _____ in.
   c) yellow stick _____ in.
   d) black stick   _____ in.
   e) nail   _____ in.
   f) clothespin   _____ in.

2. Write a ratio for each of the following:

   a) The length of the yellow stick to the length of the black stick _____

   b) The length of the nail to the length of the clothespin
   _____

   c) The length of a table to its width _____

   d) Your height (inches) to your weight (pounds) _____

   e) The length of the red stick to the length of the yellow stick
   _____

   f) The length of the green stick to the length of the black stick
   _____

   g) The length of the yellow stick to the length of the green stick
   _____

   h) The length of the classroom to the width of the classroom
   _____

3. Pat's desk is 12 ice-cream sticks long and 8 ice-cream sticks wide. Circle the ratios that can be used to compare its length with its width.

a) (8 : 12)   c) (6 : 4)   e) (2 : 3)
b) (12 : 8)   d) (3 : 2)   f) (4 : 1)

4. Write a ratio for each statement.

a) A wall had 5 feet of width for every 2 feet of height. _____
b) A dog was winning a tug-of-war with a boy, since there were 10 pounds of dog for every 8 pounds of boy. _____
c) A very thin man weighs 130 pounds and is 74 inches tall.

_____

5. Refer to the line segments below. Measure each segment using the cm. ruler. Let each letter represent the measure of the line segment. Complete the following to indicate the ratios of these measures.

a) (2 : 3) = (f : _____)
b) (5 : 2) = (_____ : _____)
c) (1 : 3) = (_____)
d) (7 : 5) = _____

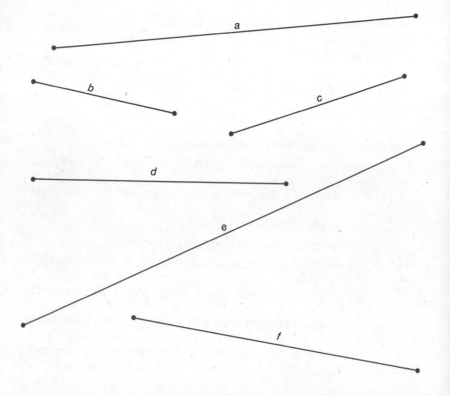

R-A-3

problem: How do you illustrate a given ratio?

materials:

10 red and 20 blue cubes
5 hinges and 25 screws
12-inch ruler

procedure:

1. Suppose we wish to illustrate a matching of a set of triangles with a set of squares that are in the ratio (3:4). To do this, we might draw the following:

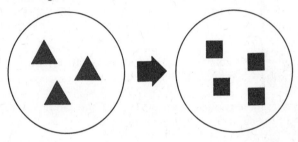

Complete the following drawing to illustrate the same ratio:

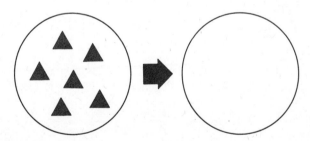

2. A ratio for the number of tables to the number of students is (1:4). How many tables are there for 28 students?_____There-fore (1:4) = (_____:28).

3. Place three red cubes and five blue cubes on the table. A ratio for the number of red cubes to the number of blue cubes is (_____:_____).

Add six red cubes for a total of nine. How many blue cubes must you have so that there are three red cubes for every five blue cubes? _____ A ratio for the number of red cubes to the number of blue cubes is (9:_____). We can say that (3:5) = (9:_____).

4. Place four hinges on the table. Place enough screws on the table so that the ratio of the number of hinges to the number of screws is (1:6). Therefore (1:6) = _____.

5. Draw line segments *AB* and *CD* so that a ratio for the length of segment *AB* to the length of segment *CD* is (2:3). If segment *AB* is four inches long, how long would *CD* be?_____

6. Draw segment *GH* so that a ratio of the length of segment *EF* (below) to the length of segment *GH* is (1:2).

E   F

Segment *EF* must have one unit of length for every_____ units of length of segment *GH*.

7. Draw segment *JK* so that a ratio for the length of segment *JK* to the length of segment *LM* (below) is (2:3).

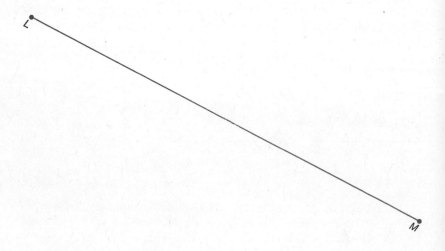

R-A-4

problem: How do you use extended ratios to compare the numbers of objects in three sets or to compare three measures?

materials:

   3 green and 3 blue sticks
   9 red, 15 green, and 21 blue cubes
   12-inch ruler
   Masking tape

procedure:

1. An extended ratio for the number of triangles to the number of squares to the number of circles, as pictured below, is (2:3:5).

Finish this drawing so that the number of triangles in the first rectangle is to the number of squares in the second rectangle is to the number of circles in the third rectangle as (2:3:5).

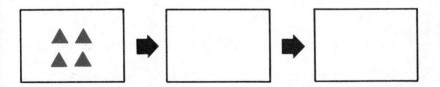

Finish this drawing so that it also shows the extended ratio (2:3:5).

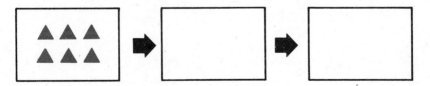

In each of the drawings above, for every two triangles there are three squares and five circles. Make another drawing, different from the first three, in which the comparison of triangles with squares with circles is (2:3:5).

2. Complete the following extended ratio to show the comparison of the numbers of red, green, and blue cubes: (3:_____:_____).

3. The three green sticks have lengths of 3 inches, 4 inches, and 5 inches. The extended ratio that compares their lengths (from shortest to longest) is (3:_____:_____). The lengths of the three blue sticks are _____ inches, _____ inches, and _____ inches. The extended ratio that compares their lengths (from shortest to longest) is (6:_____:_____).

4. Tape the green sticks together (as in fig. 4) to form a triangle. Do the same with the blue sticks. Do you notice anything about the two triangles?

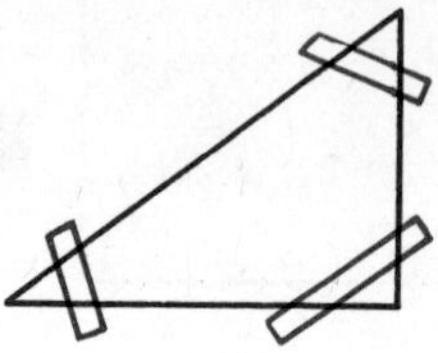

Fig. 4

5. Dave plans to cut three red sticks so that their lengths can be compared by the extended ratio (3:4:5). If he cuts the shortest stick 9 inches long, how long should he make the other two?_____ and _____.

6. Sam Gravelcement makes concrete by mixing cement, sand, and gravel in the extended ratio (1:2:3). This means that for every measure of cement he uses_____ measures of sand and _____ measures of gravel. He has just placed four buckets of cement in a cement mixer. How many buckets of sand must he add?_____ how many buckets of gravel?_____ Complete this proportionality so that it describes Sam's mixture: (1:2:3) = (4:_____:_____).

R-A-5

problem: How do you write equal ratios and extended ratios?

materials:

    18 triangles
    12 small squares
    16 large squares
    4 hinges
    24 screws

procedure:

1. Place eight triangles and twelve small squares on the table. What is the ratio of the number of triangles to the number of squares?

2. Place the eight triangles and twelve squares in two equal piles on the table. (One pile should contain the same number of squares and the same number of triangles as the other pile.) For every four triangles there are_____squares. This ratio can be written_____.

3. Next make four equal piles of the eight triangles and twelve squares. In each pile place two triangles and_____squares. For every two triangles there are_____squares. This ratio can be written_____.

4. You should have written three different ratios. Each ratio represents the same comparison of the number of triangles with the number of squares. Therefore these ratios are equal. That is, $(8:$_____$) =$ (_____$:6) =$ _____. An equation relating two ratios is called a proportion.

5. Place sixteen large squares on the table. Pretend they are sandwiches to be divided among four hungry students. A ratio for the number of sandwiches to the number of students is_____. Divide the sandwiches evenly among the four students. There are_____ sandwiches for each student. This ratio can be written (_____$:1$). So $(16:4) = ($_____$:1)$.

6. Place one hinge and six screws on the table. Six screws are needed for each hinge. The ratio of the number of screws to the number of hinges must be $(6:$_____$)$. Place three more hinges on the table with as many screws as are needed. Since we will need_____ screws for the four hinges, $(6:1) = ($_____$:4)$.

7. Place eight triangles, twelve small squares, and sixteen large squares in a pile on the table. The extended ratio that compares triangles with small squares with large squares is (_____$:12:$_____$)$.

Now make two equal piles. Place four triangles,_____small squares, and eight large squares in each pile. The extended ratio that compares the numbers of triangles, small squares, and large squares in each pile is (_____:_____:8).

Finally make four equal piles. Place _____triangles,_____ small squares, and_____large squares in each pile. The extended ratio representing the comparison of triangles with small squares with large squares in each pile is_____.

You should have written three different extended ratios. They all represent the same comparison, so they are equal to one another. That is,_____ = _____ = _____. An equation relating a pair of extended ratios is a proportionality.

8. Supply the numbers missing from each ratio (or extended ratio).

   a) $(7:3) = ($_____$:9)$
   b) $($_____$:2) = (20:4)$
   c) $(1:12) = (5:$_____$)$
   d) $(1:2:3) = ($_____$:$_____$:21)$
   e) $($_____$:15:17) = (16:$_____$:34)$
   f) $($_____$:1) = (10:2)$
   g) $(8:24:26) = ($_____$:12:$_____$)$

R-A-6

problem: How do you use a measuring wheel?

materials:

   50-foot tape
   Measuring wheel
   Stakes

procedure:

   1. Roll the measuring wheel through two turns. Measure the distance the wheel traveled in making two full turns. Write a ratio for the number of turns of the wheel to the number of feet it traveled.

   2. Now roll the wheel in a straight line for twelve turns. The ratio of the number of turns to the number of feet traveled is $(12:\square)$. (Find a number to replace $\square$.) Your ratio in the first problem was _____. So _____ $= (12:\square)$. What must replace $\square$? _____

   3. Use the tape to measure the distance for twelve turns of the wheel. _____ feet. How does this compare with your answer in exercise 2?

   4. How many turns must you roll the wheel to measure 100 feet? $(2:5) = (\square:100)$ In each ratio the first number is the number of _____. The second number is the number of _____. What is your replacement for $\square$? _____

   5. Measure a distance of 100 feet. Roll your wheel. How many turns did it take? _____. How does your answer compare with your replacement for $\square$ in exercise 4?

   6. Using your measuring wheel, place two stakes in the ground 50 yards apart. These are the starting and finishing positions for the 50-yard dash. Measure the distance between the stakes with the tape. _____ How does this answer compare with the answer you got in exercise 5? _____

   7. Roll the wheel for twenty-five turns. What should the distance be? _____ Check the distance with the tape.

   8. Measure a distance (say, 75 feet) along some curved path around the school. Ask another member of your team to measure the same path. Compare your results. _____

   9. How many turns will the wheel make in going a mile? _____

R-A-7

problem: How can you use ratios to measure speed?

materials:

    50-foot tape
    Measuring wheel
    Stopwatch
    Bicycle
    2 stakes

procedure:

1. Measure the distance you can walk in ten seconds. Write a ratio for the number of feet walked to the number of seconds of walking. (_____ : 10)

2. If you continued walking at this rate, you would be going _____ feet every 10 seconds. How far would you go in 60 seconds? (_____ : 10) = (□ : 60) (Find a replacement for □.) In these ratios the first term is the number of feet and the second term is the number of _____.

3. Walk in a straight line at normal speed for 60 seconds. Use the measuring wheel to measure the distance you walked. _____

4. Wally Walker walked 50 feet in 10 seconds. His ratio of the number of feet traveled to the number of seconds of traveling is (50 : 10) or (5 : 1). We say that Wally's walking speed is 5 feet per second, or 5 fps. How many seconds would it take Wally to walk one mile (5280 feet) at this speed? This is the same as_____ minutes and_____ seconds.

5. Ratios are frequently used to show speed. The second term of the ratio is usually 1.

When comparing the number of feet traveled with the number of seconds of traveling—

    a) the ratio (4 : 1) means 4 feet per second;
    b) the ratio (6:1) means_____; and
    c) the ratio_____ means 9 feet per second.

6. Roger Bannister was the first person to run one mile in less than four minutes. If he could have run for one hour at the rate of four minutes per mile, how many miles could he have run? His average speed was_____ miles per hour (mph). One mile is_____ feet. Four minutes is_____ seconds. Therefore the ratio of the

number of feet traveled to the number of seconds of traveling is
$(5280:240)$. $(5280:240) = (\square:1)$. What is the correct replacement for
$\square$? _____Therefore we know that_____fps is the same as
15 mph.

7. Can you use the information and procedures you learned in
exercises 1-6 to find how fast you ride a bicycle? If so, write the rate
of travel in fps and then change it to mph.

R-A-8

problem: What is a ratio compass and how do you use it?

materials:

    Ratio compass
    Drawing board
    Meter stick
    Worksheet R-A-8

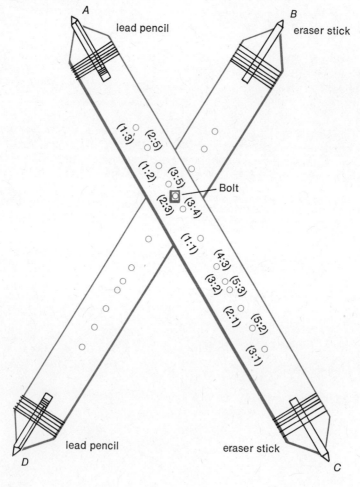

procedure:

    1. Place the bolt in the hole marked (2:3) on the ratio compass and tighten the nut on the bolt. The compass now has two short arms

and two long arms. Open the compass until the ends of the short arms (labeled *A* and *B* in the figure) are 10 centimeters apart. Measure the distance between the ends of the longer arms (labeled *C* and *D* in the figure). _____ cm. Draw line segments *AB* and *CD* and complete the following: (length of segment *AB* : length of segment *CD*) = (2:3) = (10:_____).

2. Complete the following chart for the (2:3) setting.

Length of segment *AB* (cm.)	10			18	
Length of segment *CD* (cm.)	15	18	21		36

For any opening of the compass, (length of segment *AB* : length of segment *CD*) = (2:3). Name the other ratios that can be set on the compass. If you have time, make a chart for each ratio like the one for (2:3).

3. Worksheet R-A-8 shows five line segments, each named *AB*. For each, use the compass to construct another line segment (name it *CD*) such that (length of segment *AB* : length of segment *CD*) = (2:3). In each case use a proportion to find the length of *CD*.

a) Length of segment *AB* is_____ cm.
(_____ : ☐) = (2:3)
☐ =_____
Length of segment *CD* is_____ cm.
Check by measuring *CD*.

b) Length of segment *AB* is_____ cm.
(_____ : ☐) = (2:3)
☐ =_____
Length of segment *CD* is_____ cm.
Check by measuring *CD*.

c) Length of segment *AB* is_____ cm.
(_____ : ☐) = _____
☐ =_____
Length of segment *CD* is_____ cm.

d) Length of segment *AB* is_____ cm.
_____ = _____
☐ =_____
Length of segment *CD* is_____ cm.

e) Length of segment *AB* is_____ cm.
_____
_____
Length of segment *CD* is_____ cm.

4. Tape the piece of poster board to the drawing board. On the poster board draw a triangle *KLM* such that side *KL* is 18 cm. long, side *LM* is 24 cm. long, and side *MK* is 30 cm. long. (There should be a 90° angle between sides *KL* and *LM*.) With the ratio compass set at (2:3), adjust the larger opening to match side *LM*. Draw a line segment equal in length to the distance between the points at the smaller opening of the compass. Label it *EF*. Now adjust the larger opening of the ratio compass to side *KL*. Draw a line segment equal in length to the distance between the points at the smaller opening of the compass. Draw this segment at point *E* so that it forms a 90° angle with segment *EF*. Label it *EG*. Adjust the larger opening of the ratio compass to side *MK*. Compare the smaller opening with the distance between points *F* and *G*.

5. Use proportions and the lengths of the sides of the larger triangle to predict the lengths of the sides of the smaller triangle.

Measure the sides of the smaller triangle. How do these lengths compare with the results you got using proportions?

6. Refer to triangle *KLM* in exercise 4. Construct a triangle *RST* so that (length of *KL* : length of *RS*) = (length of *KM* : length of *RT*) = (length of *LM* : length of *ST*) = (2:3). Compare triangles *KLM* and *RST*.

7. Use proportions and the lengths of the sides of triangle *KLM* to predict the measures of the sides of triangle *RST* of exercise 6.

Measure the sides of the triangle *RST*. How do these measurements compare with your predictions?

R-A-9

problem: If two figures have the same shape, how do the lengths of their sides compare?

materials:

Envelope containing 2 cardboard triangles and 2 cardboard quadrilaterals
Meter stick
12-inch ruler
Protractor

procedure:

1. Refer to the triangles in the envelope. What do you notice about their shapes?

The triangles should be matched in the following way:	Draw lines to match the sides:

Vertices

P   B
Q   Z
R   K

Sides	
PQ	BZ
QR	ZK
RP	BK

Measure the sides of the triangles:

PQ _____ cm.   BZ _____ cm.
QR _____ cm.   ZK _____ cm.
RP _____ cm.   BK _____ cm.

Write the ratios of the pairs of matching sides:

(length of side PQ : length of side KB) = _____
(length of side QR : length of side BZ) = _____
(length of side RP : length of side ZK) = _____
Are the ratios you just wrote equal to each other? _____
If they are all equal, what is the simplest ratio equal to all three? _____

2. Refer to exercise 1. Using the lengths of sides PQ, QR, and RP and the ratio you found, find the lengths of sides KB, BZ, and ZK. Are these the same measures that you wrote down in exercise 1? If not, check your work.

3. Refer to the two triangles. What do you notice about the angles at corresponding vertices of the two triangles?

Definition ► If two triangles are the same shape, the following are true:
   a) Matching pairs of angles are the same size.
   b) The ratios of the measures of the matching sides are equal.

---

4. Refer to the cardboard figures *KLMN* and *CRSA*. The matching vertices are

The lengths of the sides of the two figures are

KL _____ cm.	*CR* _____ cm.
LM _____ cm.	*RS* _____ cm.
MN _____ cm.	*SA* _____ cm.
NK _____ cm.	*AC* _____ cm.

Connect the names of the matching sides.

KL	*CR*
LM	*RS*
MN	*SA*
NK	*AC*

Are the matching angles the same size? _____

Are the matching sides the same length? _____ Which of these gives the ratio of the pairs of matching sides?

   (1 : 2)   (4 : 13)   (1 : 3)   (5 : 13)

Definition ► If two four-sided figures are the same shape, the following are true:
   a) Matching pairs of angles are the same size.
   b) The ratios of the measures of the matching sides are equal.

R-A-10

problem: How many paper clips are in the mystery envelopes?

materials:

Balance
Mystery envelopes A, B, and C
Large paper clips
Lead weights
Empty envelope

procedure:

1. Adjust the scale so that the tip of the pointer is at mark +. Place lead weights on one side of the balance and paper clips on the other. What are the smallest numbers of lead weights and paper clips that will balance each other (lead weights on one side of the balance and paper clips on the other)? Express your result as a ratio: (number of lead weights : number of paper clips).

2. Use the ratio you determined in exercise 1 to predict the number of paper clips in each mystery envelope. (Remember to use the empty envelope.)

Envelope A has _____ clips.
Envelope B has _____ clips.
Envelope C has _____ clips.

3. Compare your results with those of other students.

R-B-1

problem: How are ratios used to make comparisons?

materials:

      Envelope containing 8 red and 12 blue chips
      Egg beater
      Balance
      Block of hardwood
      Block of plastic foam
      Large paper clips

procedure:

    1. In the envelope there are eight red and _____ blue chips. The ratio of the number of red chips to the number of blue chips can be written as (8:12).

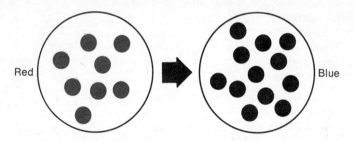

    One student shows that there are four red chips for every six blue chips. He uses the ratio (4:6). A second student prefers to match two red chips with _____ blue chips. His ratio is (2:____). The set of red chips can be compared with the set of blue chips by any of these ratios: (8:12), (4:____), (____:3).

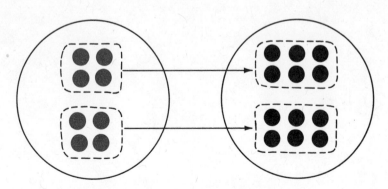

First student's comparison

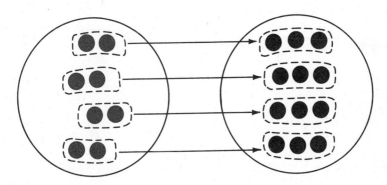

Second student's comparison

The ratio (2:3) means—

    a) There are _____ red chips for every _____ blue
       chips.
    b) There are two-thirds as many red chips as _____ chips.
    c) For every _____ red chips there are _____ blue
       chips.

2. Turn the handle of the egg beater. While the handle makes one turn, the beater makes _____ turns. While the handle makes two turns, the beater makes _____ turns. We can use the ratio _____ to show that for every turn of the handle the beater makes _____ turns. The ratio (2:____) could also be used.

3. The blocks of hardwood and plastic foam are the same size. Adjust the scale and then find how many clips it takes to balance each block. The ratio of the weight of the hardwood in clips to the weight of the plastic foam in clips is _____. This ratio has these meanings: for every _____ clips that the block of hardwood weighs, a block of plastic foam of the same size weighs _____ clips; the hardwood weighs about _____ times as much as the plastic foam.

R-B-2

problem: How can ratios help to predict events?

materials:

Thumbtack
Solid with 4 colored faces (each face an equilateral triangle)
Pair of dice

procedure:

1. When you toss a thumbtack, it will land in either of two ways:

Toss a thumbtack twenty times. How many times did it land in position A? in position B? Guess the number of times it will land each way if you toss it forty times. Toss the tack twenty more times and total your results for forty tosses.

	A	B
20 tosses		
Guess for 40 tosses		
Results for 40 tosses		

2. The object with the four colored faces can land with any of the faces down.

Face that is down	red	yellow	green	blue
Number of tosses				

Throw the solid fifty times. Chart the number of times it falls with each color down. For a single throw, which color is most likely to be down? _____

What is your best guess for the number of times *green* would be down in 100 throws? _____

3. Toss a pair of dice fifty times and complete the following table.

	Sum of numbers on upper faces of the two dice										
	2	3	4	5	6	7	8	9	10	11	12
Number of times the given sum was obtained in 50 rolls											
Guess of the number of times the given sum would be obtained in 100 rolls											
Number of times the given sum was obtained in 100 rolls											

How did you make your guesses and how good were they?_____

R-B-3

problem: What are the gear ratios of a bicycle?

materials:

Bicycle with 3 gears (masking tape on rear wheel)

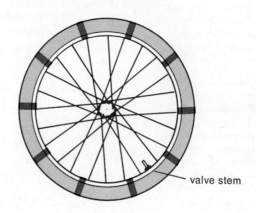

valve stem

procedure:

1. Pieces of tape have been placed at equal intervals around the rear wheel of the bicycle (see figure). The wheel is divided into _____ parts.

2. Turn the bicycle upside down. Find the number of turns of the rear wheel for each turn of the foot pedal. Do this for each gear position and fill in the first column of the chart below. Start with the valve stem down and a foot pedal up. Turn the foot pedal slowly. Don't let the wheel coast. Show the number of turns of the rear wheel to the nearest tenth of a turn.

(number turns of rear wheel : number turns of foot pedal)

Gear	For one turn	For two turns	For three turns
High	(_____ : 1)	(_____ : 2)	(_____ : 3)
Middle	(_____ : 1)	(_____ : 2)	(_____ : 3)
Low	(_____ : 1)	(_____ : 2)	(_____ : 3)

When the bicycle is in low gear, the rear wheel makes _____ turns for each turn of a foot pedal.

In which gear would you travel farthest for one turn of a foot pedal? Why? _____

In which gear might you get the greatest speed? Why? _____

---

3. Find the numbers of turns of the rear wheel for two turns of the foot pedal. Show your findings in the chart for exercise 1. Remember: do not let the bicycle coast.

4. Guess the numbers of turns of the rear wheel for three turns of the foot pedal and put the guesses in the chart for exercise 1. Check your guesses.

5. What are the advantages of having gears on a bicycle?_____

R-B-4

problem: When are two ratios equal?

materials:

Meter stick with inch markings
Worksheet R-B-4

procedure:

1. Measure segments *AB, CD,* and *EF* in inches (to the nearest inch) and in centimeters (to the nearest millimeter). Record these measurements in the following table. Predict the length of segment *GH* in centimeters.

RATIO

	Length in inches	Length in centimeters
AB	(_____ :	12.7)
CD	(_____ :	_____)
EF	(_____ :	_____)
GH	(40 :	_____)

2. The ratio (5:12.7) means that for every 5 inches of length there are 12.7 centimeters of length.

If the terms of (5:12.7) are doubled, we get (_____:_____).
If the terms of (5:12.7) are tripled, we get (_____:_____).

3. Wilt Stilton is mixing concrete for his basketball area. He is using the extended ratio (1:2:3) to represent his mixture of cement, sand, and gravel. This means that he mixes one part cement,_____ parts sand, and _____ parts gravel.

Wilt plans to put four shovelfuls of cement in the mixer. How many shovelfuls of sand and of gravel must he use?

$$(1:2:3) = (4:\_\_\_\_:\_\_\_\_)$$

(1:2:3) means that for each shovelful of cement he uses, he must use _____ shovelfuls of sand and _____ of gravel. How can the missing numbers in the extended ratio above be found? _____

4. There are six pizzas for four people, so there are three pizzas for two people, or $1\frac{1}{2}$ pizzas for each person. Complete the following

ratios. Each can be used to compare the number of pizzas with the number of people.

$$(6:\underline{\hspace{1cm}}) \qquad (\underline{\hspace{0.5cm}}:\underline{\hspace{0.5cm}}) \qquad (1\tfrac{1}{2}:\underline{\hspace{1cm}})$$

The three ratios are equal. The second ratio can be obtained by multiplying each term of $(6:4)$ by $\frac{1}{2}$. If we multiply each term of $(6:4)$ by $\frac{1}{4}$, we get $(1\tfrac{1}{2}:\underline{\hspace{1cm}})$.

Definition ▶ If k is any positive number, $(a:b) = (ka:kb)$ and $(a:b:c) = (ka:kb:kc)$.

---

5. Fill in the missing numbers.

a) $(3:8) = (9:\underline{\hspace{1cm}})$
b) $(5:3) = (\underline{\hspace{1cm}}:1)$
c) $(20:5) = (\underline{\hspace{1cm}}:1)$
d) $(3:4:5) = (12:\underline{\hspace{0.8cm}}:\underline{\hspace{0.8cm}})$
e) $(8:8) = (1:\underline{\hspace{1cm}})$

R-B-5

problem: Do equal ratios have any special properties?

materials:

Worksheet R-B-4
Meter stick

procedure:

1. Is the proportion $(9:2) = (18:4)$ true? Why, or why not?

_____

In $(9:2) = (18:4)$, the 9 and the 4 are called the *extremes* and the 2 and the 18 are called the *means*.

Find the product of the extremes. _____

Find the product of the means. _____

How do the product of the extremes and the product of the means compare? _____

2. Find the product of the extremes and the product of the means for each of the following proportions.

Proportion	Product of the extremes	Product of the means
$(30:5) = (6:1)$		
$(6:4) = (1\frac{1}{2}:1)$		
$(4\frac{1}{2}:2) = (9:4)$		
$(5:12.7) = (10:25.4)$		

How do the two products compare in each case? _____

> *Product rule* ► In a true proportion the product of the extremes equals the product of the means.

3. Follow the rule to write each proportion as an equation involving products.

a) $(2:3) = (8:12)$ $2 \times 12 = 3 \times$ _____

b) $(4:2) = (2:1)$ $4 \times$ _____ $=$ _____ $\times 2$

c) $(2:3) = \left(1:1\frac{1}{2}\right)$ $2 \times$ _____ $=$ _____ $\times$ _____

d) $(3:8) = (6:16)$ _____ × _____ = _____ × _____
e) $(4:10) = (10:25)$ _____
f) $(54:10) = (\square:1)$ _____
g) $(32:\square) = (64:4)$ _____

4. Solve each proportion for $\square$.

   a) $(4:\square) = (2:6)$ *d)* $(3:5) = (\square:2\frac{1}{2})$

   b) $(4:12) = (10:\square)$ *e)* $(\square:24) = (3:4)$

   c) $(120:2\frac{1}{2}) = (\square:1)$ *f)* $(50:7.39) = (\square:23.42)$

R-B-6

problem: How can you use a bicycle wheel to measure distance?

materials:

  50-foot tape, marked in tenths of a foot
  Bicycle wheel mounted on a fork
  4 stakes
  Chalk
  Hammer
  String

procedure:

1. Fix a stake in the ground. Starting at the stake, roll the wheel in a straight line. Measure (to the nearest tenth of a foot) the distance the wheel traveled in making ten turns. _____ feet. What is the ratio of the number of turns of the wheel to the number of feet traveled? _____

2. Fix two stakes (*A* and *B*) in the ground about 70 paces apart. Use the bicycle wheel and a proportion to find the distance from *A* to *B*. (Remember: roll the wheel in a straight line.) Do this twice to check for accuracy. Then set up the following proportion:

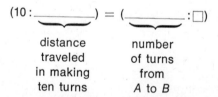

$$(10 : \underline{\hspace{2cm}}) = (\underline{\hspace{2cm}} : \square)$$

  distance   number
  traveled   of turns
  in making   from
  ten turns   *A* to *B*

Find the correct replacement for $\square$. What does this represent?

Use the tape to measure the distance from *A* to *B*. _____ feet. How do your answers compare? _____

(If your two answers differ by more than $\frac{1}{2}$ foot, try again.)

3. Now mark off 100 paces with two stakes. Measure the distance, using the bicycle wheel and complete the following proportion.

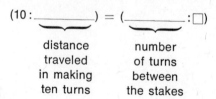

$$(10 : \underline{\hspace{2cm}}) = (\underline{\hspace{2cm}} : \square)$$

  distance   number
  traveled   of turns
  in making   between
  ten turns   the stakes

4. Cut a piece of string 25 feet long and use it to draw a circle with a radius of 25 feet. Make sure the string is taut and straight at all times. Measure the distance around this circle, using your bicycle wheel. Distance: _____.

5. Measure the distance along some path. Make the measurement twice. Compare your results with those of your classmates.

R-B-7

problem: How do you measure distances on a globe?

materials:

Globe of the earth
String

procedure:

1. Locate the equator on the globe. The earth's equator is 25,000 miles long and contains 360 arc degrees. The curved lines that come together at the poles are called meridians. Find the meridian through Greenwich, England (near London). Locate the point where this meridian crosses the equator. This will be our zero point. From here we will measure distances along the equator. The equator is marked off (in each direction) in arc degrees. Find where the equator passes through the Isle of Celebes (in the East Indies). It is _____ arc degrees from our zero point. Therefore (_____ : ☐) = (360 : 25,000). So ☐ = _____ (the distance in miles from our zero point to the Isle of Celebes).

2. How far is it from Chicago, Illinois, to Cairo, Egypt? The shortest path can be found by stretching a string from Chicago to Cairo on the globe. Can you think of a way of finding the number of arc degrees in the path? This path contains _____ arc degrees. Complete this proportion. Then solve it to find the distance from Chicago to Cairo. (360 : 25,000) = (_____ : ☐)

3. Jacksonville, Florida, and Shanghai, China, are at about the same latitude (31° N). Find the shortest path from Jacksonville to Shanghai. If you flew along this path, would you go east or west of St. Louis, Missouri? _____ Would your flight take you north of the Arctic Circle? _____ What would be the length of your flight in arc degrees? _____ in miles? _____ How far would you fly if you traveled westward along the 31° N latitude line from Jacksonville to Shanghai? _____

4. How far is it from Buenos Aires, Argentina, to Paris, France? _____ arc degrees; _____ miles

R-B-8

problem: How can you measure distances on a road map?

materials:

Local road map
Map distance wheel
Meter stick

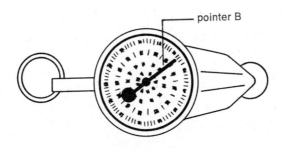

map distance wheel

procedure:

1. The map distance wheel measures distance along any path. Set the pointer B at 0. Roll the wheel along a line segment that is 100 centimeters long. How far does pointer B show that the wheel has moved?

2. Select four locations (such as towns) on your road map. Measure the distance between each pair of locations (along highways), using the distance wheel. Use proportions and the scale on the map to find the actual distance (in miles) between pairs of locations. Make a chart of your work.

R-B-9

problem: What do ratios have to do with the enlargement of figures?

materials:

Cardboard polygons (*PQRS* and *ABCD*)
Meter stick
Drawings of a dog
Sheet of tracing paper

procedure:

1. Do the cardboard polygons have the same shape? Compare each angle of polygon *ABCD* with each angle of polygon *PQRS*. Draw lines to show the pairs of angles that have the same measure.

∠*DAB*	∠*SPQ*
∠*CBA*	∠*PQR*
∠*DCB*	∠*SRQ*
∠*CDA*	∠*PSR*

2. Draw lines to show the pairs of corresponding sides and diagonals. Measure each side and each diagonal, and complete the following table.

First polygon	Second polygon	Segment	Length
AB	*PQ*	*AB*	
BC	*QR*	*BC*	
CD	*RS*	*CD*	
DA	*SP*	*DA*	
AC	*PR*	*AC*	
BD	*QS*	*BD*	
		PQ	
		QR	
		RS	
		SP	
		PR	
		QS	

3. Using the measurements from the table above, complete the following ratios.

(length of segment *AB* : length of segment *SP*) =   (18:12) =   (3:2)
(length of segment *BC* : length of segment *PQ*) = _____ = _____
(length of segment *CD* : length of segment *QR*) = _____ = _____
(length of segment *DA* : length of segment *RS*) = _____ = _____
(length of segment *AC* : length of segment *SQ*) = _____ = _____
(length of segment *BD* : length of segment *PR*) = _____ = _____

4. There are two drawings of a dog. One is an enlargement of the other. Do the dogs in the drawings have the same shape? _____ Measure each of the following segments.

Segment	Length	Segment	Length
BC	10.0 cm.	EF	_____ cm.
AC	_____ cm.	DF	_____ cm.
AB	_____ cm.	DE	_____ cm.

Complete the following ratios:

Ratio

(length of segment BC : length of segment EF) = (10 : ____) = (1 : ____)
(length of segment AC : length of segment DF) = (____) = (____)
(length of segment AB : length of segment DE) = (____) = (____)

5. Put a sheet of tracing paper over the larger drawing and copy points D, E, and F. Cut out triangle DEF.

Does ∠E have the same measure as ∠B? _____
Does ∠D have the same measure as ∠A? _____
Does ∠F have the same measure as ∠C? _____

6. If two figures have the same shape, it appears that the following statements are true:

a) Corresponding angles have _____.
b) Corresponding sides and diagonals are _____.

R-B-10

problem: How do you make and use a rubber-band enlarger?

materials:

    4 rubber bands
    Drawing board
    Large sheet of paper
    Meter stick
    Sheet of tracing paper
    Large drawing of a dog

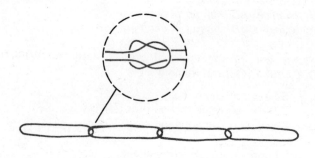

Fig. 1

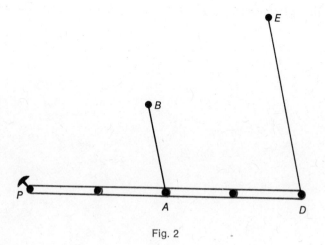

Fig. 2

procedure:

1. Place a piece of paper on the drawing board and draw line segment *AB* 16.0 centimeters long. Make a rubber-band chain from four
rubber bands looped together as shown in figure 1. Fasten one end of

the chain to the drawing board very securely. (See fig. 2.) Put a pencil in the other end of the chain. Stretch until the middle knot is at point A. Mark the position of the pencil D. Locate E in the same way. Draw segment DE. The ratio of the measure of segment AB to the measure of segment DE should be _____. Use this ratio to find the length of segment DE. _____ Measure segment DE. _____ How does this answer compare with the one you got when you used ratios?

2. Mark a point F to the right of segment DE so that D, E, and F form the vertices of a triangle. Using a sharp pencil, stretch the rubber-band chain to F. Mark the point at the middle knot of the chain. Label it C.

Draw triangles DEF and ABC.
Measure segment DE. _____ cm.
Measure segment EF. _____ cm.
Using the fact that corresponding sides of the two triangles have lengths in the ratio (1:2), find the length of segment AC and the length of segment BC.
Check your results.

3. Copy triangle ABC on tracing paper. Mark points A, B, and C on this triangle. Place the sheet over triangle DEF to see whether the corresponding angles of triangle ABC and triangle DEF are the same size. What did you find? _____

4. Make an enlarged copy of the picture of the dog. Be careful. (Hint ► Keep your eye on the middle knot.)

R-B-11

problem: How do you use a mirror to find the height of an object?

materials:

Mirror (about 6″ × 8″)
Marble or level (for leveling the mirror)
String and weight (for plumb line)
Felt-tip pen
Yardstick, marked in tenths of a foot

procedure:

1. Select an object whose height you would like to find. (The area around its base must be level.) Mark a point in the middle of the mirror with the felt-tip pen. Label the point *M*. Lay the mirror flat on the ground (or floor) and level it.

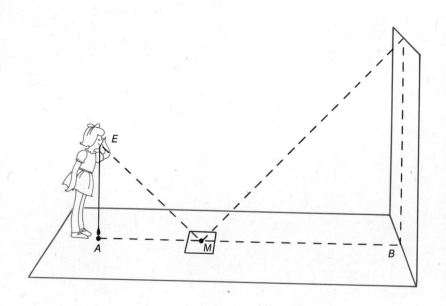

2. Stand where you can do both of the following (see the figure):

   a) Lean slightly forward and hold a plumb line to your eye (*E*) so that it touches the floor at a point (*A*).
   b) See the top of the object at *M* on the mirror.

3. Have a partner mark point *A* and a point *B* at the base of the object you are measuring.

Measure segment *EA*. ————————————————

Measure segment *MA*. ————————————————

Measure segment *MB*. ————————————————

4. How are triangle *EAM* and triangle *CBM* related? Use a proportion and the results of exercise 3 to find the length of *BC*.

5. Repeat the experiment, placing the mirror a different distance from the foot of the object. How do your two results compare? ————

———————————————————————————————————

6. Measure the height of another object, using the same procedure.

R-B-12

problem: How is speed measured?

materials:

Stakes
Stopwatch
Bicycle
Measuring wheel
Car

procedure:

1. Measure a distance of 300 feet. Have a partner time you as you ride the bicycle over this distance. _____ seconds. Using the ratio (feet traveled : seconds of traveling), do the following:

    a) Write a proportion relating this ratio to the ratio ($\Box$:1).
    b) Solve the proportion for $\Box$.
        $\Box$ = _____ .   Your speed was _____ feet per second (fps).

2. Twenty-two feet per second (fps) is the same as 15 miles per hour (mph). Therefore (fps:mph) = (22:15). Use this fact and your results from exercise 1 to find your speed in miles per hour. _____ mph.

3. Measure a distance of 400 feet along a road. Have someone drive this distance at a rate of 20 miles per hour. Ride in the car, and start the stopwatch as you reach the start of the run. Stop it when you finish.

    _____ seconds of traveling
What was your rate in feet per second? _____
What was your rate in miles per hour? _____

4. Find the average speed in fps and mph represented by each of the following pieces of data.

Time of traveling	Distance traveled	Speed	
		fps	mph
10 sec.	300 ft.		
12 sec.	550 ft.		
8.0 sec.	580 ft.		
8.0 sec.	150 ft.		

5. How many seconds does it take to travel 500 feet on a bicycle at a speed of 25 mph? _____

R-B-13

problem: What is *compression ratio*?

materials:

    Manuals for cars
    Model of a piston and cylinder
    Metric ruler
    Scale marked in tenths of an inch
    Glue

procedure:

    1. Move the piston up and down in the cylinder. The nail keeps the piston from moving to the top of the cylinder.

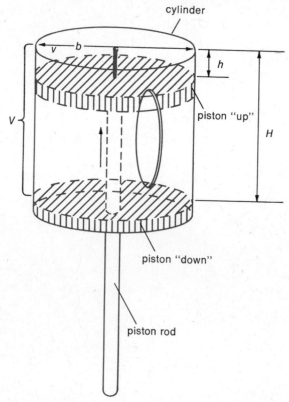

When the piston is all the way down, the volume of the gasoline-air mixture in the cylinder is V. When the piston moves all the way up, the gasoline-air mixture is compressed to a volume v. At this instant the spark plug fires. The mixture burns and the expanding gases push the piston down. The piston is connected to a drive shaft by the piston rod in such a way that the downward thrust turns the drive shaft and moves the car.

Find the following measures to the nearest tenth of an inch (see the figure).

$h =$ _____in.
$H =$ _____in.
$b =$ _____in. *b* represents *bore* (the inside diameter of the cylinder).

2. $(H : h)$ is the compression ratio of the model. Find this ratio and solve the following proportion for □. $(H : h) = (□ : 1)$. □ $=$ _____.

3. $H - h$ is the distance the piston travels. The movement through the given distance is called the *stroke* of the piston. The length of the stroke is _____ inches.

4. $V =$ _____ cubic inches. $v =$ _____ cubic inches.

(Hint ► Volume of a cylinder equals π times the square of the radius of the base times the height. In this case use $\dfrac{b}{2}$ for the radius, h and H respectively for the two heights, and 3.14 for π.) $V - v$ is called the *piston displacement*. Find $V - v$. _____

5. *Summary.* Our "cylinder" has a compression ratio of (_____ : 1), a bore of _____ in., a piston stroke of _____ in., and a piston displacement of _____ cu. in.

6. In a 1969 Ford 250 six-cylinder engine—

bore $= 3.68$ in.   stroke $= 3.91$ in. $v = 5.48$ cu. in.

Find the piston displacement for this engine. The piston displacement is _____ cu. in. for each cylinder. The displacement for all six cylinders is _____ cu. in. The volume V of each cylinder is _____ cu. in. The ratio $(V : v) = (_____ : 5.48) = (_____ : 1)$.

7. The piston displacement for a Honda 90 is 90 cubic centimeters. Using the conversion ratio (cubic inches : cubic centimeters) $=$ (cu. in. : cc.) $= (1 : 16.4)$, find the number of cubic inches in the piston displacement. _____ cu. in.

8. *Optional.* Find the bore and stroke of a piston in a car manual. Compute the piston displacement for each cylinder. _____ cu. in.

R-B-14

problem: How can you measure the speed of a pitched baseball?

materials:

Baseball and mitt
50-foot tape
Weight
9-foot pole
Chair
Board, 2′ × 2′ square

procedure:

1. Mark off a distance of 60 feet 6 inches. This is the distance from the pitcher's mound to the catcher. To measure the time it takes a ball to travel this distance involves measuring the distance a weight will fall during the time the ball is in motion.

Mark off the pole in tenths of a foot. Place the board on the ground and place the pole upright beside the board. Stand on the chair, facing the pitcher and catcher, and hold a weight at some arbitrary point beside the pole. Have two observers stand midway between you and the catcher. You are to drop the weight the instant the ball leaves the pitcher's hand. Keep dropping the weight from different heights until everyone is satisfied that the catcher catches the ball at the same time that the weight strikes the board. The observers should hear the sounds of both at the same time.

2. Use the chart on the following page to find the time the ball was in flight.

For example, a weight will drop 8.3 feet in about_____seconds.

3. The ball is usually released $3\frac{1}{2}$ feet in front of the pitcher's mound, so it travels 57 feet to the catcher. If this takes .72 seconds, what is the ball's speed in feet per second? _____ in miles per hour? _____

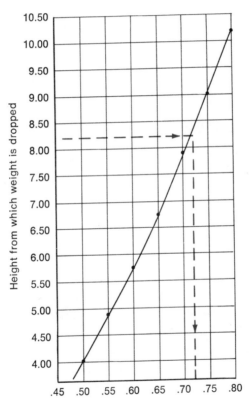

Seconds weight takes to fall (at sea level)

**Time It Takes a Weight to Drop
from a Given Height at Sea Level**

Teacher's Guide

The answers that appear in the Teacher's Guide for ratio are those that may be difficult to arrive at. Answers are provided only when slight variations in the activity will not affect them. A set of guidesheets on which answers pertinent to the equipment to be used are supplied may prove helpful.

R-A-1

Students are introduced to the use of ratios. They begin by comparing two sets of objects by two different methods: first, element to element and then set to set. In the example at the beginning the concept of *proportion* is introduced. Though this term is not used in the guidesheet, more advanced students may be helped by having it pointed out to them.

R-A-2

The student applies what he has learned about ratios to compare lengths of objects. It could be pointed out to him that this is basically the same type of activity that was given in guidesheet R-A-1, with this exception: he is comparing numbers of units of measurement rather than the cardinalities of two sets.

Give each student one red stick 3 inches long; one green stick 6 inches long; one yellow stick 12 inches long; and one black stick 8 inches long.

R-A-3

On guidesheets R-A-1 and R-A-2 the students were shown sets of objects or sets of measurements and were asked to write ratios to compare the sets. On guidesheet R-A-3 the problem is reversed: they are given ratios and asked to complete sets of objects to illustrate the ratios.

In exercise 6 segment *EF* is 3 inches long; segment *GH* should be drawn 6 inches long. In exercise 7 line segment *LM* is 6 inches long; segment *JK* should be drawn 4 inches long.

R-A-4

Care should be taken in this activity that the term *ratio* is not used. A ratio is a comparison of the cardinalities of two sets (or of two measures); we will call a comparison of the cardinalities of more than two sets (or of more than two measures) an *extended ratio*, though this is not common usage. Only extended ratios are involved in this guidesheet.

Give each student three green sticks (3 inches, 4 inches, and 5 inches) and three blue sticks (6 inches, 8 inches, and 10 inches).

Students may have some difficulty reading exercise 1 if they have not developed the habit of reading $(a:b) = (c:d)$ as "a is to b as c is to d" and, similarly, of reading $(a:b:c) = (d:e:f)$ as "a is to b is to c as d is to e is to f."

For exercise 4 a right triangle can be formed from the set of green sticks or from the set of blue sticks.

R-A-5

The students were introduced to equal ratios in earlier activities. In R-A-3 and R-A-4 they were asked to complete sets of objects to keep ratios equal. In R-A-5 the activity is carried further and they are asked to complete a ratio with no clue other than another ratio to which it is equal. To do this, they must work with a constant of proportionality. It is not necessary for them to know this terminology at this time. For example, in $(7:3) = (\underline{\quad}:9)$ (exercise 8a), 3 is multiplied by 3 to get 9, and therefore 7 must be multiplied by 3 to arrive at 21, which will keep the ratios equal; it is possible for the student to do this without recognizing the number 3 as a constant of proportionality.

R-A-6

Students are introduced to a practical application of ratio—measuring curved paths with a measuring wheel.

See guidesheet F-1a, Functions, at the end of chapter 3, for detailed instructions for constructing a measuring wheel. All measures for that wheel were in metric units. For activity R-A-6 we are not assuming that the students are able to work with anything other than whole numbers. This guidesheet is written for use with a wheel that has a circumference of 2.5 feet so the student can use the ratio $(2:5)$ (see exercise 4). Therefore it is important that construction for the wheel be done carefully so that its diameter has a measure very close to 9.55 inches.

In exercise 8 the students must measure the same path in order to compare their results. The first student to measure the path should mark off his route in some way so that the others can follow.

R-A-7

Students are shown another common application of ratio: the use of ratio to measure speed.

The measuring wheel should be carefully constructed (see the Teacher's Guide for R-A-6, above). A stopwatch is needed for this activity. If the math classroom or laboratory does not have one, it might be possible to borrow one from the physical education department. Stopwatches can be purchased for about $13 and up.

Students should work in teams of at least three members. To make the statement in exercise 5 more convincing, call the students' attention to the familiar terms *mph* and *rpm*. If a car is going 60 mph, what ratio is described? (60:1) If a record is played at a speed of 45 *rpm,* what ratio is described? (45:1)

The activity in exercise 7—measuring the speed of a bicycle—is an outdoor activity. There are similar speeds that the students could investigate. For example, how fast can they roller-skate or ice-skate? How many feet per second can they swim?

It will be necessary for the students to make approximate measurements when they work with counting numbers only. Since students have at least some intuitive knowledge of fractions, however, this may be a good place to teach something about this topic.

R-A-8

Students learn about the ratio compass and are shown how it can be used to simplify drawing figures in proportion.

Ratio compasses can be purchased through school supply stores or they can be made from two flat, pointed sticks, each about 20 inches long, appropriately drilled and bolted together. Properties of similar triangles are the basis of their construction. Settings on a ratio compass are (1:3), (1:2), (2:3), (1:1), (3:4), (3:5), (2:5), (5:2), (5:3), (4:3), (3:2), (2:1), and (3:1).

The worksheet should show segments with lengths 16 cm., 20 cm., 22 cm., 10 cm., and 18 cm.

The activity can be extended by varying the ratio setting on the compass and by having the students draw additional line segments.

The triangles that the students construct in exercise 4 should be smaller than the first triangle they drew and similar to it. The triangles that they construct for exercise 6 should be larger than the first triangle and similar to it.

R-A-9

Students use ratios to discover that if two figures have the same shape, regardless of size, corresponding sides can be compared by equal ratios.

Each student should be supplied with an envelope containing two cardboard triangles and two cardboard quadrilaterals: triangle *PQR* with sides *PQ* = 18 cm., *QR* = 16 cm., and *RP* = 10 cm.; triangle *KBZ* with sides *KB* = 27 cm., *BZ* = 24 cm., and *ZK* = 15 cm.; quadrilateral *KLMN* with sides *NM* = 6 cm., *ML* = 8 cm., *LK* = 7 cm., and *KN* = 5 cm.; and quadrilateral *CRSA* with sides *CR* = 18 cm., *AC* = 24 cm., *SA* = 21 cm., and *RS* = 15 cm.

In exercise 3 the students should notice that the triangles are the same shape, and they should describe the comparison of the lengths of their sides with the ratio (2:3). In exercise 6 they should select the ratio (1:3) to describe the comparison of the two quadrilaterals.

R-A-10

Students learn how they can use ratios to determine the quantity of an item in an envelope by comparing its weight with the weight of a known quantity of the same item.

A simple balance can be made from two pieces of soft wood, such as pine, for the base and upright, a piece of pressed board for a pointer, a piece of pegboard for the arm, two milk cartons, a nail, and three paper clips (two to attach the milk cartons and one to adjust the balance). See the illustration. Inexpensive balance scales can be purchased instead, and will not change the procedure of the activity. Fishing sinkers and curtain weights are examples of readily available lead weights. The lead weights should all weigh the same.

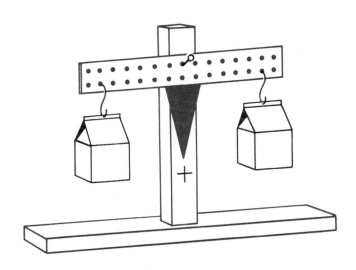

"Mystery envelopes" should be prepared in advance and a list made of the number of clips in each. Make sure a reasonably small whole number of lead weights balances a whole number of paper clips. The empty envelope must be identical to the mystery envelopes. An empty envelope must be placed on the balance opposite the mystery envelope to counterbalance the weight of the envelope.

R-B-1

Ratios can be used to compare numbers of objects or events, as well as units of measurements.

Poker chips or small pieces of colored poster board can be used for the red and blue chips. You may want to ask several students to bring egg beaters from home, or possibly the home economics department could provide them. The wooden block should be hardwood and the same size as the block of plastic foam. Small nails or other common small objects of uniform size and weight can be substituted for the large paper clips.

In exercise 2 the students should arrive at a ratio of about one to four. This will vary, depending on the egg beater used.

R-B-2

The guidesheet presents a group of three activities that demonstrate the use of ratios to make predictions.

A colored tetrahedron can be made from the pattern shown below.

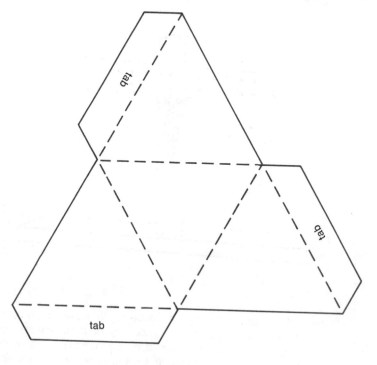

The faces can be colored with paint or crayon or covered with colored paper (note that the chart in exercise 5 refers to the sides as red, yellow, green, and blue) and the tabs can be fastened with glue or tape. A similar activity can be done with a spinner that has four positions, but students usually prefer rolling a tetrahedron. A feltboard to toss the objects on will keep the noise level down. Students may benefit from working together to find the answers to the questions.

R-B-3

An active learning situation that boys particularly will enjoy applies what the students know about ratios to a comparison of the number of turns of the rear wheel of a bicycle per turn of the pedal for the different gears of a three-speed bicycle.

Turn the bicycle upside down and mark off ten equal arcs on the rear wheel (see the picture on the guidesheet). This can be done with chalk or masking tape. One mark should differ from the others in some way to help students keep track of the number of full turns of the wheel.

The chart below shows results from one three-speed bike. Students' results may differ slightly from these.

Gear Ratios

Gear	For One Turn	For Two Turns	For Three Turns
High	(3.4:1)	(6.8:2)	(10.2:3)
Middle	(2.6:1)	(5.2:2)	(7.8:3)
Low	(1.9:1)	(3.8:2)	(5.7:3)

R-B-4

The purpose of the activity is to help students discover the definition that precedes exercise 5: If k is any positive real number, $(a:b) = (ka:kb)$ and $(a:b:c) = (ka:kb:kc)$. The term *extended ratio* is introduced here. Students should realize that this is not a standard term. It is used here to point out the close relation between ratios (comparisons of two sets) and comparisons of more than two sets, which are more commonly referred to simply as comparisons.

The worksheet should show three line segments, AB, CD, and EF, 5 inches, 10 inches, and 15 inches long respectively.

Since there are 12.7 cm. in 5 inches, there are 25.4 in 10 inches, 38.1 in 15 inches, and 101.6 in 40 inches.

R-B-5

The terms *proportion, means,* and *extremes* are defined, and the students are led to discover the multiplication theorem: In a true proportion the product of the means equals the product of the extremes.

They will again work with worksheet R-B-4, on which are drawn line segments AB, CD, and EF.

R-B-6

Practice is provided in using proportions to determine an unknown distance by measuring it with a wheel of unknown circumference.

A plywood handle for the bicycle wheel can be made according to the directions for the measuring wheel in guidesheet F-1a in chapter 3; the measurements will have to be adjusted to accommodate the larger wheel.

You may want to vary the instructions for this activity by fixing the stakes in the ground yourself before class. The first two should be about 200 feet apart; call these A and B. The second two should be about 300 feet apart; call these C and D. The circumference of a circle with a 25-foot radius is approximately $157\frac{1}{7}$ feet. For exercise 5 have students choose some curved path and mark it off in some way so that they can measure this same path again to check their work.

R-B-7

An interesting application of ratio is presented. Given that the circumference of the earth at the equator is about 25,000 miles and that there are 360 degrees in a circle, it is possible to approximate any great-circle distance on the globe simply by measuring this distance with a string, determining how many arc degrees the string would cover on the equator, and setting up a proportion.

The distance from Jacksonville, Florida, to Shanghai, China, by the shortest (great circle) route is about 8000 miles (to the nearest hundred miles). The distance from Jacksonville to Shanghai, flying westward along the 31° N latitude line, is approximately 9400 miles (to the nearest hundred miles). The distance from Chicago, Illinois, to Cairo, Egypt, is about 6100 miles (to the nearest hundred miles).

To vary and extend this activity, you might have students suggest points on the globe that they would be interested in finding. It is usually interesting to find how far from your home are cities in other parts of the world that are currently in the news. For example: How far are you from Moscow? from Paris? from Saigon? from Seoul, Korea? Cairo, Egypt? How far out of its way does a jet travel if it is hijacked to Cuba while flying from Los Angeles to New York City? A guessing game built on this activity in which students guess a distance first, and then measure it, will increase interest still further.

R-B-8

A practical and interesting application of ratio is presented: measuring distance on a map, using a small wheel called a map distance wheel that can be obtained from a science supply company.

The students might enjoy extending this activity by using a road atlas to plan an imaginary car trip. They can use the map distance wheel to determine which is the best route to take between two cities by answering questions such as these: Is it better to take a stretch of road that is 350 miles long on which you can average 40 miles an hour or to take a road that is 500 miles long on which you can average 65 miles an hour? If you are driving from Portland, Maine, to Detroit, Michigan, and would like to plan a side trip to Montreal, how many extra driving hours must you allow for, assuming that you will average 55 miles an hour? Once students have decided on a route, they can determine how far they could hope to drive each day and determine where they would want to reserve hotels.

An interesting point may be brought up: Aren't relative distances on a map distorted? The answer is yes, for certain types of maps, but on most road maps this distortion is negligible.

R-B-9

The activity is an exercise in using ratios to compare two figures of the same shape but of different size.

Polygon *ABCD* should have the following measures: $AB = 18.0$ cm., $BC = 7.5$ cm., $CD = 15.6$ cm., $DA = 12.5$ cm. Polygon *PQRS* should have the following measures: $SP = 12.0$ cm., $PQ = 5.0$ cm., $QR = 10.4$ cm., $RS = 8.4$ cm.

The picture to be enlarged should be a simple drawing, a cartoon figure, or something you have drawn yourself. The illustration below will work nicely. One picture should be a twice size enlargement of the other, and segment *BC* should be 10.0 cm. long.

Guidesheets R-A-8 and R-A-9 fit well between R-B-9 and R-B-10, although you may want to make the exercises in those guidesheets more challenging for use at this point.

R-B-10

The students are shown how to enlarge a simple drawing, using what they learned about similar figures in the preceding activity. They will probably want to extend this activity by working with additional pictures.

If drawing boards cannot be borrowed from the art department, perhaps students can make them. If not, drawing boards can be purchased from any art supply store. Something must be inserted in the board that will hold the rubber-band chain very securely and not fly out when the rubber-band chain is pulled taut. A cup hook or flat-headed screw or nail would work best; a thumbtack can also be used, but care must be taken that it is secure.

R-B-11

The students will probably want to extend the activity by measuring objects other than those suggested.

A string weighted with a sinker or other relatively heavy object will make a good plumb line.

Encourage the students to do this activity with great care. The mirror must be level and even with the base of the object being measured. They should use a fine unit of measurement: millimeters, sixteenths of an inch, in tenths and hundredths of a foot. If the activity is carefully done, results will be impressively accurate.

R-B-12

The students use proportions to translate from one set of units to another in measuring speed.

See guidesheet R-A-7, Teacher's Guide, for suggestions regarding the stopwatch and guidesheet F-1a in chapter 3 for instructions for making measuring wheels. The students should be able to supply bicycles.

If it is not possible to carry out the activity with the car as suggested in exercise 3, the work can be done by substituting the following information: it takes approximately 13.64 seconds to travel 400 feet at 20 miles per hour. Similar problems can be set up from the information in the chart above. The horizontal scale shows time in seconds, and the vertical scale shows distance in feet. The slanting lines represent speed in miles per hour. For example, if you travel 550 feet in 7.5 seconds (see point A), your speed is 50 mph; if you travel 350 feet in 8.5 seconds (see point B), your speed is about 28 mph.

For exercise 4 the students should obtain answers close to the following:

a) 30.0 fps; 20.5 mph
b) 45.8 fps; 31.3 mph
c) 72.5 fps; 49.4 mph
d) 18.8 fps; 12.8 mph

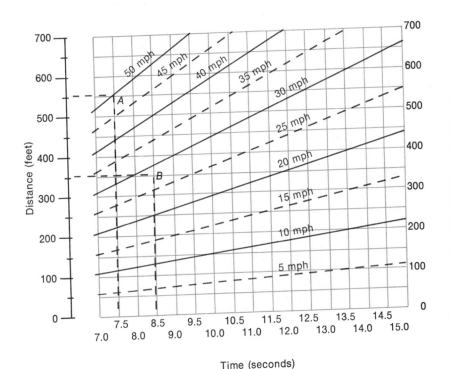

Average Speeds: Converting fps to mph (e.g. moving 350 feet in 8.5 seconds gives an average speed of 28 mph.)

R-B-13

The activity will be of particular interest to boys; it allows them to apply extracurricular knowledge to their work with ratio. They may be able to find information in motorcycle and race car magazines as well as in car manuals. It should also appeal to some girls, since it provides insight into a topic about which they may know very little.

The model of the piston and cylinder can be made from an empty oatmeal box, a circle of plywood (cut to fit the box), a long nail, and a

$\frac{3}{4}$-inch dowel rod. Fasten the dowel to the center of the piece of plywood with the nail so that about $\frac{3}{4}$ inch of the nail is above the plywood. Cut a hole slightly more than $\frac{3}{4}$ inch in diameter in the center of
the base of the box. Fit the plywood into the box so that the dowel
passes through the hole. Cut the box to the proper height (so that the
piston can move anywhere from 3 to 4 inches) and glue the box top
onto the box. Now cut a viewing hole in the side of the box. (See the
figure.)

The bore of a cylinder ranges from about 3 inches to about $4\frac{1}{2}$
inches and the stroke of a piston ranges from about 3 inches to about
4 inches. Following are approximate statistics of the 1969 Ford 250
six-cylinder engine:

Piston displacement (1 cylinder):   41.57 cu. in.
                    (6 cylinders): 249.42 cu. in.
Volume of each cylinder: $V = 47.05$ cu. in.
Ratio $(V:v)$ or $(H:h):(47.05:5.48) = (8.58:1)$

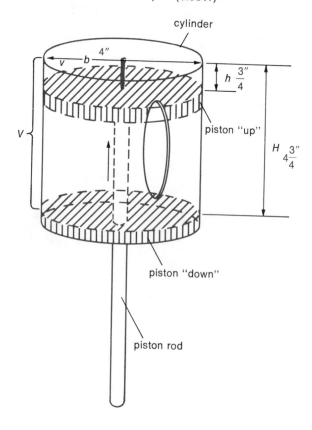

R-B-14

The students are shown how to measure the speed of a pitched baseball. This project demands a great deal of accuracy, so students must concentrate on their work. They will want to discuss and experiment with various ways of dropping the weight so that it leaves the hand of the person dropping it at the same time the ball leaves the pitcher's hand.

The tape should be marked off in tenths of a foot and attached after the pole has been put in the ground. The weight can be any compact object that can be held and released easily, will be minimally affected by air resistance, and will make a fairly loud sound on striking the board. When students have agreed upon a speed, they might want to compare this speed with that of other fast-moving objects.

Professional baseball pitchers throw the ball at speeds of between 80 and 100 miles per hour.

Bibliography

Black, Millard H. "Characteristics of the Culturally Disadvantaged Child." In *The Disadvantaged Child: Issues and Innovations,* edited by Joe Frost and Glenn Hawkes. Boston: Houghton Mifflin, 1966.

Bloom, Benjamin S., et al., eds. *Taxonomy of Educational Objectives, Handbook I: Cognitive Domain.* New York: McKay, 1956.

Bruner, Jerome S. "Some Theorems on Instruction Illustrated with Reference to Mathematics." In *Theories of Learning and Instruction,* edited by Ernest R. Hilgard. Sixty-third Yearbook of the National Society for the Study of Education, pt. I. Chicago: the Society, 1964.

_____. *Toward a Theory of Instruction.* New York: Norton, 1968.

Dienes, Z. P. *The Power of Mathematics.* London: Hutchinson Educational, 1964.

Dorrie, Heinrich. *100 Great Problems of Elementary Mathematics.* New York: Dover, 1965.

Gardner, Martin. *The Second Scientific American Book of Mathematical Puzzles and Diversions.* New York: Simon & Schuster, 1961.

_____. "Of Sprouts and Brussels Sprouts: Games with a Topological Flavor." *Scientific American* 217 (July 1967):112–115.

Hill, Winfred F. *Learning: A Survey of Psychological Interpretations.* San Francisco: Chandler, 1963.

Johnson, George. *Education for the Slow Learners.* Englewood Cliffs, N. J.: Prentice-Hall, 1963.

Krathwohl, David R., et al., eds. *Taxonomy of Educational Objectives: The Classification of Educational Goals,* Handbook II: *Affective Domain.* New York: McKay, 1964.

Montessori, Maria. *The Montessori Method.* Introduction by J. McV. Hunt. New York: Schocken, 1964.

National Council of Teachers of Mathematics. *The Learning of Mathematics, Its Theory and Practice.* Twenty-first Yearbook. Washington: the Council, 1953.

Niven, Ivan, and Zuckerman, H. S. "Lattice Points and Polygonal Area." *American Mathematical Monthly* 74 (December 1967):1195–1200.

Nuffield Mathematics Project. *I Do, and I Understand.* New York: Wiley, 1967.

Polya, George. *Mathematical Discovery: On Understanding, Learning, and Teaching Problem Solving.* Vols. I and II. New York: Wiley, 1962.

Purkey, W. W. *The Search for Self: Evaluating Student Self-Concepts.* Gainesville: Florida Educational Research and Development Council, 1968.

Riessman, Frank. "The Overlooked Positives of Disadvantaged Groups." In *The Disadvantaged Child: Issues and Innovations,* edited by Joe Frost and Glenn Hawkes. Boston: Houghton Mifflin, 1966.

Small, Dwain, et al. *The Problem of Under Achievement and Low Achievement in Mathematics Education.* Report of Project No. H-307. Washington: Department of Health, Education, and Welfare, 1966.

Wolff, Peter H. *Developmental Psychologies of Jean Piaget and Psychoanalysis.* New York: International Universities Press, 1960.

Woodby, Lauren. *The Low Achiever in Mathematics.* Washington: Government Printing Office, 1965.

Index

Other Professional Books in Education
Published by
Science Research Associates, Inc.

Benson, Charles S. *The School and the Economic System.* 1966.

Chesler, Mark, and Fox, Robert. *Role-Playing Methods in the Classroom.* 1966.

Engelmann, Siegfried. *Preventing Failure in the Primary Grades.* 1969.

Fox, Robert; Luszki, Margaret; and Schmuck, Richard. *Diagnosing Classroom Learning Environments.* 1966.

Gottlieb, David, and Ramsey, Charles. *Understanding Children of Poverty.* 1967.

Heddens, James. *Today's Mathematics.* 1964.

Hillson, Maurice, and Bongo, Joseph. *Continuous Progress Nongraded Education: Inventions, Innovations, and Implementations.* 1970.

Ianni, Francis. *Culture, System, and Behavior.* 1967.

Joyce, Bruce R. *Strategies for Elementary Social Science Education.* 1965.

_____ and Harootunian, Berj. *The Structure of Teaching.* 1967.

Lazarsfeld, Paul, and Henry, Neil. *Readings in Mathematical Social Science.* 1966.

Miller, Robert. *Statistical Concepts and Applications.* 1968.

_____ *Tests and the Selection Process.* 1966.

Robinson, H. Alan, and Rauch, Sidney. *Guiding the Reading Program: A Reading Consultant's Handbook.* 1965.

Schmuck, Richard; Chesler, Mark; and Lippitt, Ronald. *Problem Solving to Improve Classroom Learning.* 1966.

Stodola, Quentin, and Stordahl, Kalmer. *Basic Educational Tests and Measurements.* 1967.